Broadsides

Caricature and the Navy 1756–1815

James Davey & Richard Johns

Contents

Seaforth
PUBLISHING

in association with

NATIONAL
MARITIME
MUSEUM

Caricature and the Navy

Johnohn Bull and Napoleon face one another across a narrow stretch of water, locked in a war of words. From the shores of 'Falsehood', Napoleon releases a flurry of fantastical reports announcing the dispersal of the British fleet and celebrating a successful invasion of England. John Bull, firmly grounded on the side of 'Truth', meets the onslaught of French propaganda with a single fanfare proclaiming the 'Total Defeat' of Britannia's enemies while holding aloft a copy of the *London Gazette Extraordinary*, the special edition of the Government newspaper, published on 6 November 1805, with the first reports of the British victory, and Nelson's death, at the Cape of Trafalgar. Meanwhile, in the distance, broadsides of a different kind are levelled at the French and Spanish fleets.

John Bull Exchanging News with the Continent first appeared in the window of Samuel Fores's Piccadilly print shop on 11 December, days after the return of Nelson's body, when the printmaker and his publisher could have been confident of an enthusiastic reception for their latest collaborative work. With its combination of patriotic humour and political urgency, George Woodward's memorable print is typical of countless caricatures produced during the French Revolutionary and Napoleonic Wars: where the character of an entire nation could be embodied by the no-nonsense figure of John Bull; and where the violent uncertainties of a global conflict could be reduced to the simple polarities of 'Truth' and 'Falsehood'. Furthermore, by highlighting the importance of newsmaking and print culture in the ongoing struggle against

George Woodward
John Bull Exchanging News
with the Continent
Samuel Fores,
11 December 1805.
Hand-coloured etching.
NMM PAF4004

James Gillray
Fighting for the Dunghill – or
– Jack Tar settling Buonaparte
Hannah Humphrey,
20 November 1798.
Hand-coloured etching with
aquatint.
NMM PAD4792

Jack Tar straddles the globe, taking the British Isles and Europe in his stride as he sends Napoleon off the edge with a bloody nose. Gillray produced another, almost identical but cruder version of *Fighting for the Dunghill*, in which 'Citoyen Francois' replaces Napoleon as the recipient of Jack's blows. Jack Tar's physical resemblance to George III is deliberate, intended as a further sign of the British sailor's loyalty in contrast to his revolutionary French foe.

Fighting for the DUNGHILL __ or __ Jack Tar settling BUONAPARTE .

Napoleon's France, Woodward's print itself becomes implicated in the war effort – a single, well-aimed shot in a continuous volley of patriotic satire.

This book explores the various ways in which the Royal Navy was represented in the art of caricature during the half century that encompasses the Seven Years War (1756–63), the American War of Independence (1775–83), and the French Revolutionary and Napoleonic Wars (1792–1815). During this period of near-continuous military engagement, the navy confirmed its status as Britain's primary defence against foreign invasion and a vital tool of an expanding empire. The dramatic growth of British commerce and wealth during the eighteenth century had been reliant on the protection given by the Royal Navy, and on the access to new routes and areas of trade that became possible as a result of its global presence. The navy was the nation's greatest expense and its largest employer; in wartime, it provided work for hundreds of thousands of people, not only on its ships, but also across a multitude of trades in its dockyards, and in the fields and forests that supplied the British fleet with enormous quantities of food and timber.

With so much at stake, the fortunes of Britain's sailors around the globe received unprecedented attention from a news-hungry populace, and the government's management of the nation's navy became a subject of intense public interest. The navy's senior commanders became household names, as admirals who found success in battle were raised to the status of national hero. Meanwhile, the seamen who served under them came to represent a more robust ideal of virtuous masculinity, embodied in the irrepressible figure of Jack Tar.

This period of heightened engagement with the navy coincided with a particularly vibrant episode in the history of graphic satire in Britain. The age of Vernon, Rodney, Howe and Nelson was also the age of James Gillray, Thomas Rowlandson and Isaac Cruikshank, all still remembered today as leading figures in the history of British caricature. The insatiable demand for contemporary satire in Georgian Britain also provided a profitable living for other printmakers, including Robert Dighton, Charles Williams and the aforementioned Woodward, artists who are less well known today but whose individual humour and distinctive style were recognised and valued in their own time. These and other printmakers, together with the publishers they worked with and for, were capable of exerting a profound influence on the behaviour and attitudes of the many thousands who encountered their work. Their subjects were many, touching on every aspect of contemporary politics, religion and society; and for all, the navy and its exploits were a recurring and urgent theme.

Caricatures of contemporary political and military affairs were routinely published within days of the events that inspired them, often appearing in tandem with, and even anticipating, the official dispatches and news reports that filled an ever-growing number of daily and weekly papers. This journalistic concern for current affairs ensured that caricature served as both a barometer of and a guiding force for public opinion of the navy. As well as focusing on specific events and personalities, the most effective caricatures could convey a bigger picture – presenting a broader, strategic vision of the nation's domestic and foreign affairs to a non-specialist audience.

The majority of naval caricatures were explicitly patriotic, foregrounding significant maritime victories, throwing scorn at the enemy, or characterising the British naval officer as a model of great leadership and the ordinary sailor as a dependable and straight-talking national treasure. Within this context, printmakers and their publishers cast themselves as loyal participants in a national struggle against Britain's enemies overseas. However, no one was exempt from the satirists' attention. Virtually immune from prosecution for sedition, libel or obscenity, caricaturists were able to probe the controversies and contradictions of their age with a freedom that was unavailable to the authors of contemporary books, political pamphlets and newspapers. Defeat, dishonour and ethically dubious actions prompted some of the most savage caricatures of the age, while even the greatest victories and most conspicuous heroes could be lampooned.

The caricaturists who rose to prominence during the latter decades of the eighteenth century were able to draw upon a rich tradition of graphic satire, the modern roots of which germinated amid the new press freedoms of the seventeenth and early eighteenth centuries. Images, often comical and frequently of a political or religious nature, became a common accompaniment of the broadsides, ballads and other forms of popular print that streamed from the capital's presses. *Lucipher's new Row-Barge, for First-Rate Passengers*, a copy of which was used as a cover image for the *Weekly Journal* in 1721, is a relatively early example of a popular image that engages with Britain's maritime identity in the broadest sense – in this case as part of a larger satire on the corruption and huge personal losses associated with the South Sea Bubble in 1720. As

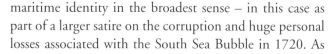

Lucipher's new Row-Barge, for First-Rate Passengers
1720.
Etching with engraving.
NMM PAH7360

The South Sea Company was founded in 1711 on the promise of untold riches that would result from a monopoly on trade in South America. However, within a decade, uncontrolled speculation in the company's stock provoked a spectacular economic crash, causing financial ruin for many. In this anonymous print, published after the company's corrupt affairs had been exposed, the devil carries away a South Sea director in one of the company's own ships, loaded with riches, as chaos and financial ruin reign all around.

the century progressed, printmakers engaged more often and directly with the navy and naval policy. The anonymous satire *Next Sculls at the Adm**ty* strikes a blow at Admiralty policy during the War of Jenkins' Ear, while Louis Boitard's *British Resentment or the French fairly Coopt* combines a dense array of text with an image that is rich in allegory and emblem to commemorate the recent siege of Louisburg and the ongoing campaign against the French in North America. By putting the faraway actions of British sailors into a global and domestic political context, each of these early satires helped to set the scene for the next generation of printmakers.

During the second quarter of the eighteenth century, William Hogarth, an artist celebrated for his penetrating social satires and innovative pictorial narratives, set a new benchmark for Georgian graphic satire. Although he was at pains to distinguish between *character*, which he regarded as his primary subject, and *caricature*, Hogarth was keenly aware of the comic and political potential of the latter. His satirical portrait of John Wilkes, one of his final works, depicts the radical politician with a sneering expression, demonic wig, and an ill-fitting 'cap of liberty' balancing preposterously

Next Sculls at the Adm**ty
7 December 1744.
Etching.
NMM PAG8517

Louis-Philippe Boitard
British Resentment or the
French fairly Coopt at
Louisbourg
Thomas and John Bowles,
25 September 1755.
Hand-coloured etching.
NMM PAF3915

above his head. It is a true caricature, distorting and exaggerating the physical appearance of a prominent political figure to devastating comic effect.

More generally, Hogarth's emphasis on the morals and manners of contemporary society, his delight in the comic potential of everyday objects, and his love of visual and verbal puns, all had a formative influence on the work of later caricaturists. In Hogarth's paired images *Beer Street* and *Gin Lane*, for example, Gillray and his contemporaries found contrasting character types that would re-emerge in altered form in their own work as the well-fed, patriotic figures of John Bull and Jack Tar, and the emaciated and depraved sans-culottes of revolutionary France.

William Hogarth
John Wilkes, Esq.
16 May 1763.
Etching with engraving.
British Museum Satires 4050

Hogarth produced this acerbic print in response to an earlier attack by Wilkes, who had taken offence at his inclusion in Hogarth's anti-war print *The Times*. The portrait, which shows Wilkes during his trial for seditious libel, sold in large numbers and was copied relentlessly. However, in the process, the image transcended its original satirical purpose to become an icon of the Wilkesite cause.

William Hogarth, Beer Street and Gin Lane
1751. Etching with engraving. British Museum Satires 3126, 3136

War and revolution abroad, and the rise of radical politics at home, provided an abundance of material for the politically aware artist. At the same time, the techniques and technology of printmaking continued to develop in ways that would profoundly affect the look of caricature. Etching surpassed line engraving (Hogarth's preferred technique) and mezzotint as a faster, more versatile method for translating an image onto copper for printing. Meanwhile, the greater availability of watercolour pigments meant that relatively simple designs could be transformed with a full range of colour, applied by hand. Images became bolder as a result: direct enough to catch the eye from a print shop window, yet sufficiently engaging to persuade the passing customer to enter and buy a print of their own. The lengthy speech bubbles and explanations that were often a necessary component of earlier satires became less prominent as artists and their public became familiar with a new, leaner pictorial language defined by an acute Hogarthian observation of politics and society and a continually evolving repertoire of recognisable characters.

At the same time, improvements in distribution and advertising, coupled with an exponential growth of newsprint and print shops in London and in major towns countrywide, meant that the audience for caricature became larger and more diverse than ever before. Hawkers and mail coaches carried caricatures to every county, and packets of new prints were routinely sent abroad where, to a foreign audience, they inevitably accrued new and unexpected meanings. During the Peace of Amiens, for example, the engraver Abraham Raimbach was surprised to find the passport office at Calais 'decorated with the masterly caricatures of Gillray, ridiculing the chief personages of the English Administration'.[1]

Caricature flourished in the fast-moving, opportunistic environment of the modern city, where political intrigue was endemic, and where gossip spilled from taverns, clubs and coffee houses into scandal magazines, newspapers and print shops. At the top end of the market, Gillray specialised in sophisticated and trendsetting satire of the highest quality, with prices to match, sold exclusively from the premises of his publisher Hannah Humphrey in London's fashionable St James's. Rudolph Ackermann's emporium on the Strand sold patriotic prints to respectable city types alongside a range of artist's materials and fashion plates. Samuel Fores combined bawdy humour with some of the most contentious political caricatures of the period at his shop in Piccadilly, while Thomas Tegg in Cheapside catered for the booming lower end of the market, producing an astonishing range of caricatures from coarse and familiar one-liners to searching personal and political satires. Fores even made portfolios of caricatures available for hire, so those unable or unwilling to buy could enjoy an evening of topical entertainment.

Acknowledging the extraordinary success of their better-known rivals, Fores and Tegg also produced cheaper, cruder versions of popular works by Gillray and others. Such prints were usually hurried to press soon after the original in order to catch the moment, and often included modifications and additional quips that played to a different audience across town. In Fores's reworking of Gillray's celebrated *John Bull taking a Luncheon*, for example, an unappetising pot of crocodile hash swings from Nelson's hook, and the 'True British Stout' that John Bull drinks in Gillray's design has been downgraded to 'Best Home Brew'. Fores's mischievous imitation appeared just a week after Gillray's popular print.

'Caricature Shops are always besieged by the public, but it is only Mrs Humphrey's shop, where Gillray's works are sold, that you will find people of high rank, good taste and intelligence.'

Johann Christian Hüttner, one of the first chroniclers of Gillray's career, writing in 1806.[2]

James Gillray

John Bull taking a Luncheon: – or – British Cooks, cramming Old Grumble-Gizzard, with Bonne-Chére

Hannah Humphrey, 24 October 1798. Hand-coloured etching.
NMM PAF3941

John Bull taking a Luncheon was one of several caricatures inspired by the news of Nelson's victory at the Battle of the Nile. Nelson appears at the head of a succession of British admirals – including Howe, Warren, Duncan, St Vincent and Bridport – each of whom had been involved in a signal naval victory during the war against revolutionary France.

The assembled heroes each serve up their respective victories to John Bull, 'Old Grumble-Gizzard', whose appetite for 'frigasees' of enemy ships is fast becoming sated. Through the window, opposition Whig politicians are shown bewailing the turn of events, fearing that they might be next in line for 'his Guts'.

John Bull Taking a Lunch – or Johnny's Purveyors pampering his Appetite with Dainties from all parts of the World
Samuel Fores,
1 November 1798.
Hand-coloured etching.
NMM PAF3940

Caricature and the Navy

George Woodward
John Bull Peeping into Brest
Piercy Roberts,
June 1803.
Hand-coloured etching.
NMM PAF3950

Woodward was one of only a handful of artists capable of rivaling Gillray's wit. *John Bull Peeping into Brest* sees a Brobdingnagian John Bull looking down with relish on the confined French fleet and the tiny figure of Napoleon, appraising them both as 'a very light breakfast'. It is a simple but brilliant echo of the culinary theme of Gillray's earlier *John Bull taking a Luncheon*.

upon my word — a very Pretty light Breakfast.

mercy on us what a monster. — he'll swallow all my ships at a mouthful, I hope he dont see me.

IOHN BULL PEEPING into BREST

The creative unruliness of the market in which Gillray and his contemporaries operated is conveyed beautifully by *Very Slippy-Weather*, Gillray's intricate and self-referential satire on the business of caricature, in which an elderly man falls unceremoniously on an icy pavement outside Humphrey's shop at 27 St James's Street, losing his wig and the contents of his pocket. The eclectic selection of caricatures on display is echoed by the gathered spectators who, oblivious to the slapstick playing out around them, stand in admiration of Gillray's work. The variety of bawdy and topical subjects in Humphrey's window – from the foibles of modern clergymen to the absurdity of ladies' fashion – also points to the satirical milieu in which naval caricature accrued meaning.

In 1818, an early compiler of Gillray's work reflected on the astonishing range and audacity of 'the myriad graphic squibs … which have since poured so abundantly from the press, and made so marked a feature in the political history of the country'. George III's reign, he added, 'may well be designated the age of caricatura'.[3] This book takes a closer look at the eighteenth-century navy through the lens of contemporary caricature, from the politics of rank to the role of the Channel Fleet as a 'sure shield' against invasion, and from the violence of the press gang to the unexpected perils of shore leave.

VERY SLIPPY-WEATHER.

James Gillray
Very Slippy-Weather
Hannah Humphrey,
10 February 1808.
Hand-coloured etching.
British Museum Satires 11100

On a wintry day, a conspicuous thermometer allows a hapless city walker to keep an eye on the temperature (though to no avail) – just as Gillray's caricatures provide a measure of the times.

THE ENGLISH LION DISMEMBER'D
Or the Voice of the Public for an enquiry into the loss of Minorca — with Adl B—g's plea before his Examiners.

The English Lion Dismember'd Or the Voice of the Public for an enquiry
into the loss of Minorca with Adl. B—g's plea before his Examiners
1756. Hand-coloured etching with engraving. NMM PAF3986

The failure of Admiral John Byng to conclusively fight the French in 1756, during the Seven Years War, was an inauspicious moment for the British navy. Byng's inaction, and the resulting fall of Minorca to France prompted outrage among a broad section of the British public, who accused the admiral of incompetence and, even worse, cowardice. During the court martial that followed, Byng was accused of betraying the nation's trust and convicted of 'failing to do his utmost' to take or destroy the enemy's ships. For a nation so reliant on its navy and a command of the sea, this was a heinous crime indeed. The loss of Minorca brought down the government led by the Duke of Newcastle; in its place William Pitt the Elder formed a ministry. Although a sympathetic president was appointed to Byng's court martial, little could be done to assuage the anger of the public, the scorn of his naval peers, and the ire of the king, George II. The disgraced admiral was found guilty and shot on the quarterdeck of his own ship on 14 March 1757.

Unfortunately for Byng, his fate was embroiled in a deeper political game as opposition forces exploited popular opinion to force a decisive response from the government of Pitt, as a result of which Byng became a scapegoat. The details of Byng's humiliation in battle, court martial and subsequent execution were updated daily in the press. Few such reports could convey the political implications of the admiral's downfall as concisely, or as cruelly, as *The English Lion Dismember'd*, published shortly after news of the loss

of Minorca had reached England. Centred on the image of an injured lion with its paw (representing Minorca) cut off, the anonymous print contemplates the meaning of such a high-profile naval defeat and highlights the navy's vacillation over Byng's trial.

While the Admiralty sit in council beneath a framed copy of 'Byng's Plea', and as politicians search for excuses, an angry mob of militiamen demand an explanation for the apparent failings of their political superiors. 'Whores and Cards, Hunting & Horse-racing are more their concern than Commerce or Glory', declares one while brandishing a flail; 'The Gentry are more concerned to preserve the Game than their Country' adds another. Meanwhile, to the left, two Frenchmen discuss which of the lion's limbs to remove next as they turn their attention to Britain's American colonies.

Byng's defeat in 1756 raised serious questions over the government's handling of the war, and of the navy's ability to advance British interests. By presenting a naval defeat as the catalyst of a far wider national crisis, the anonymous creators of this biting image direct the force of their satire towards the administration of the day as much as at Byng himself. The sight of a dismembered English lion looking on impotently as a French cockerel pecks at a British ensign was evidently designed to cause maximum embarrassment, and may even have contributed to Byng's conviction and eventual execution. In the meantime, the admiral himself is conspicuous by his absence from the scene.

CHAPTER I

The Admirals

Since the time of Sir Francis Drake and the Armada, there has been a long history in the British Isles of celebrating naval heroes. Admirals who had honoured their rank and advanced the nation's imperial and trading ambitions by leading the navy to victory in battle were rewarded with great riches and the peculiar distinction (rare in an age before mass media) of being recognisable to a large section of society, thanks to the circulation of news reports, biographical sketches and, above all, printed portraits. As the eighteenth century progressed, the individual achievements of Britain's admirals also became a regular subject of caricature. When Gillray gave comic expression to a new canon of British heroes in *John Bull taking a Luncheon* (see page 8), Admirals Howe, Nelson, St Vincent et al, joined the likes of Blake, Shovell, Benbow and Vernon in a line of naval champions spanning the previous two centuries. Increasingly however, it was scandal – often muddied by party politics – rather than success that attracted the attention of caricaturists. Those that failed to live up to the expectations placed upon them could expect to be pilloried; the slightest accusation of corruption, abuse of privilege or (more likely and most damning of all) of failing to engage the enemy, was met with the closest scrutiny from a ruthless and fickle, but always inventive press.

Isaac Cruikshank
The Ghost of Byng
Samuel Fores,
28 March 1808.
Hand-coloured etching.
NMM PAG8598

William Hogarth and Charles Grignion
Mr Garrick in the Character of Richard III
20 June 1746.
Etching with engraving.
British Museum, Ee,3.121

Printed by W. Hogarth.   Mr. Garrick in the Character of Richard the 3d   Engraved by W. Hogarth & C. Grignion

Half a century after his execution, Byng's memory was revived by Isaac Cruikshank in response to another military scandal. In *The Ghost of Byng*, the disturbing decomposing figure of the admiral, conspicuous in an old-fashioned uniform that even in 1808 signalled a bygone era, stands before Lieutenant General Whitelocke, an army officer court-martialled after failing to capture Buenos Aires (an action depicted in the framed image behind him). Whitelocke was cashiered for his troubles – a lenient punishment, suggests Byng, who returns from the grave to remind the world of his own cruel fate.

Cruikshank's depiction of the startled Whitelocke recalls Hogarth's earlier painting and print of the celebrated Shakespearean actor David Garrick as Richard III, but the overall subject was also inspired by a pamphlet that appeared two years after Byng's execution. Presented as a tragi-comic conversation, the short text sees Byng's ghost appear to defend his honour to another army officer, Lord George Sackville, who was himself court-martialled after refusing an order for a cavalry charge during the British victory at the Battle of Minden in 1759. Their encounter, as retold by the pamphlet's anonymous author, was timely and to the point. When asked by Sackville why he had not pursued the French fleet, Byng replies defiantly that to beat an enemy 'o'er and o'er again' would be a 'paltry Action, past Forgiveness'. Towards the end of the poem, Sackville's thoughts turn towards his own impending trial (after which he too was cashiered), and to his public fate in the hands of the 'filthy Mob' with their 'damn'd Lampoons and Satires':

> [Will] Printsellers and Gravers join
> To maul this Character of mine,
> And vilely stick me up and down,
> In ev'ry Shop about the Town,
> The Butt of ev'ry gaping Clown;
> What must I say, what must I do?[4]

Byng's reply is simple, if tinged with regret: pay the scurrilous writers and printmakers to work for you instead of your enemies.

COUNT DE GRASSE *delivering his Sword to the Gallant* ADMIRAL RODNEY.

William Hamilton
The French Admiral Count De Grasse, Delivering his
Sword to Admiral (now Lord) Rodney
1782. Etching with engraving.
NMM PAD5388

Count De Grasse delivering his Sword to the Gallant Admiral Rodney
P. Mitchell,
27 May 1782.
Hand-coloured etching.
NMM PAF3711

In the spring of 1782, Britain's faltering efforts to defeat the rebellious
colonialists in North America were temporarily boosted by Admiral
Rodney's decisive victory over Comte De Grasse, admiral of the
French fleet at the Battle of the Saintes. Although it would prove to
be a rare British victory in a war that was being lost elsewhere, news
of Rodney's Caribbean victory was greeted at home by public
rejoicing and the popular refrain 'Rodney for Ever!'.

De Grasse's flagship the *Ville de Paris*, one of four French ships
captured during the battle, provides the location for an ostensibly
gentle satire on the manner of Rodney's victory, in which the
opposing admirals appear as physically contrasting but in all other
ways equal opponents. 'You have fought me handsomely,' De Grasse
declares with a bow as he offers his sword in defeat; 'I was glad of the
opportunity,' replies Rodney, as sailors from both sides 'huzza' and
toast the victorious British admiral.

In an age when the formality of naval battle and rules of
engagement were faithfully observed, and protocols of surrender and
parole meant that a defeated admiral could take comfort in refined
conversation and a shared bottle of wine with his erstwhile enemy,
Count De Grasse delivering his Sword also subtly confronts the notion
that opposing officers had more in common with each other than
with those among the lower ranks of their own side. By mocking the
dull, patriotic prints that emerged after the battle, Rodney's comic
exchange with De Grasse, who had led the French fleet to victory at
the Battle of the Chesapeake the previous year, appears at once
triumphant and absurd.

Who's in fault? (Nobody) a view off Ushant
William Humphrey,
1 December 1779.
Hand-coloured etching.
British Museum Satires 5570

This anonymous printmaker holds no punches by depicting Keppel as a 'nobody' following his failure to defeat the French at Ushant in 1778. If 'it' has a heart, the inscription goes on to suggest, 'it must lay in its Breeches'. Keppel's failure intensified political divisions at home and polarised attitudes to the war. Another high-profile court martial of a naval commander (this time resulting in acquittal) was followed swiftly by Keppel's resignation.

The Ville de Paris, Sailing for Jamaica, or Rodney Triumphant
Thomas Colley,
1 June 1782.
Hand-coloured etching.
British Museum Satires 5993

Not everyone envisaged Rodney's triumph at the Battle of the Saintes as such a gentlemanly affair. *The Ville de Paris* depicts a grotesque De Grasse on his hands and knees pulling a boat of French prisoners of war through the water towards Jamaica. On his back, a triumphant Rodney tugs at the defeated Frenchman's ponytail and prods his sword at a broken and 'discolour'd' Bourbon flag as a makeshift British flag flies proudly from De Grasse's improvised mizzenmast. Crude in its message and its technique, Colley's print was produced, perhaps, as a hurried riposte to Mitchell's overly polite caricature of recent events in the West Indies.

James Gillray
Rodney introducing
De Grasse
Hannah Humphrey,
7 June 1782.
Hand-coloured etching.
NMM PAF3710

RODNEY introducing DE GRASSE.

Rodney and De Grasse reappear in the first of a group of early political sketches produced by James Gillray and Hannah Humphrey. Rodney kneels in deference before introducing the tall and thin figure of De Grasse to a rotund George III. Standing to the King's right, Charles James Fox laments Rodney's success: 'this Fellow must be recalled,' he complains to the King, 'he fights too well for us – & I have obligations to Pigot, for he has lost 17,000 at my Faro Bank'. On George's left, Admiral Keppel recalls his own failed attempt to capture the *Ville de Paris* during the indecisive First Battle of Ushant in 1778. Against this backdrop of political discord and infighting within the navy, Gillray seems keen to promote Rodney as a patriotic figure – a commander who plays by the rules.

James Gillray
Rodney Invested – or –
Admiral Pig on a Cruize
1782.
Hand-coloured etching.
NMM PAF4156

RODNEY Invested – or – Admiral PIG on a Cruize.

Two weeks before the Battle of the Saintes, a new government restored Keppel to office as First Lord of the Admiralty. One of his first decisions was to remove Rodney as commander of the fleet in favour of Admiral Hugh Pigot; swapping a national hero for a commander who lacked experience and who was regarded by many as incompetent and corrupt. When news of Rodney's victory reached English shores, Keppel was forced to congratulate him but it was too late to recall the order. The *Morning Herald* encapsulated the apparent subterfuge by suggesting that Pigot's promotion had been engineered as a means of reclaiming his considerable gambling debts. The contrast in character and reputation between Rodney and Pigot is not lost on Gillray: as 'generous gallant Rodney' steps ashore to be greeted by Britannia and Neptune, a British lion tears at a French standard beneath the admiral's feet. Meanwhile, Pigot the 'pig' spies ashore from a ship that has playing cards for sails as Fox looks on in the distance, still clutching Pigot's IOU.

SEA AMUSEMENT.

OR COMMANDERS IN CHIEF OF CUP AND BALL ON A CRUISE,

Thomas Rowlandson

Sea Amusement. Or Commanders in Chief of Cup and Ball on a Cruise

1785.

Hand-coloured etching.

NMM PAF3713

In a ship's cabin two admirals occupy their time gambling over a child's toy. As a pile of coins and banknotes grows in front of them the proper objects of their attention – a navigational chart and the plan of a coastal fortification (alluding to contemporary proposals to reinforce the fortifications at Plymouth and Portsmouth) – lie neglected underfoot. Although the characters and details of Rowlandson's disapproving caricature may be intended as a generic comment on the inactivity of the British fleet following the loss of America, it is likely that a contemporary audience would have recognised the figure on the left as the Duke of Cumberland, brother of George III, who despite a lack of experience at sea rose rapidly to the rank of vice admiral. On the right sits Sir Edmund Affleck, a distinguished veteran of the Battle of the Saintes who was appointed rear admiral in 1784, the year before the publication of Rowlandson's print. Either way, Rowlandson's image of a decadent navy, unfit for service, is far from flattering and aimed squarely at the officers in charge. Although first produced following Britain's defeat in the War of American Independence, *Sea Amusement* was later reissued by Samuel Fores in 1802, during the Peace of Amiens, with the revised title *Commanders Engaged at Sea*.

WHAT A CUR'TIS !

Done from an Original Drawing by a British Officer — & publish'd as a Guide to Preferment

James Gillray

What a Cur'tis!

Hannah Humphrey, 9 June 1795.

Hand-coloured etching.

NMM PAF4151

Zounds, these damn'd hail stones hinder one from doing ones duty! – I cannot see-out of my Eyes for them! – ah! it was just such another cursed peppering as this, that I fell inn with on the coast of America in the last War; – what a Deuce of a thing it is, that whenever I'm just going to play the divil, I am either hinderd by these confounded French storms, or else, loose my way in a Fog. –

J.º GY; des.ª et fec'.   A French Hail Storm, — or — Neptune loosing sight of the Brest Fleet —

Pub. Dec.ʳ 10ᵗʰ 1793, by H. Humphrey N.18. Old Bond Street

James Gillray, A French Hail Storm, – or – Neptune losing sight of the Brest Fleet
Hannah Humphrey, 10 December 1793. Hand-coloured etching. NMM PAF3926

When Admiral Howe was made commander of the Channel Fleet in February 1793, he pursued a controversial policy of open blockade, deploying smaller ships to patrol the Channel within sight of French harbours, while the main body of the fleet remained in port, ready to mobilise if necessary. Although this proved a safer option for British ships and sailors, particularly during the harsh winter months, it proved ineffective as a means of blockade, allowing French ships easier passage in and out of port. As Admiral Byng had discovered to his cost a generation before, naval commanders who were perceived as failing to do their duty could expect little public sympathy. Critics of Howe's reactive tactics represented the admiral as an untrustworthy, even treasonous figure dodging in and out of Torbay in an effort to avoid a confrontation. *A French Hail Storm* offers a rare visual take on the situation, representing Howe with some irony as a Neptune-like figure, complete with a gold sea chariot drawn by a pair of dolphins.

As he heads back towards the safety of Torbay, Howe covers his eyes to protect them from a 'hail storm' of gold coins spat out by French cherubim wearing the revolutionary *bonnet rouge*. 'Zounds, these damn'd hail stones hinder one from doing ones duty!' Howe exclaims unconvincingly as the French fleet return to Brest harbour unhindered. Gillray's vision is much clearer, it seems, as he implies that Howe has been blinded by French bribes (notice that many coins happen to fall into Howe's large coat pockets), but such harsh criticism may have been unfair as Howe stood to make even more money by capturing French prizes.

The following year, Howe silenced his critics by commanding the British fleet to a signal victory and taking seven French ships at the Battle of the Glorious First of June – an event commemorated by Gillray in *John Bull taking a Luncheon*, in which the admiral is shown as one of a line of British naval heroes. After the battle it emerged that Howe, under the advice of his Captain Sir Roger Curtis, had chosen not to chase the remnants of the retreating enemy, costing the British further prizes. In 1795 Gillray took a final shot at Howe with *What a Cur'tis!*, in which Curtis appears as Howe's dog, licking the boots of 'Black Dick' (the Admiral's nickname) as he ponders the dispatches from the battle.

Charles Williams
Sternhold and Hopkins at Sea or a Slave out of Time
Samuel Fores,
August 1809.
Hand-coloured etching.
NMM PAF3945

After the French fleet had been run aground by Lord Cochrane during the Battle of the Basque Roads in 1809, the commander of the Channel Fleet, Admiral Gambier, chose to blockade the stranded enemy rather than mount a further attack – a decision that Cochrane immediately criticised, accusing his fellow officer of cowardice. Gambier insisted on a court martial to clear his name, but like Byng and Howe before him, his reluctance to pursue the enemy left a lasting mark on his reputation.

Charles Williams's complex satire, published at the height of the Napoleonic Wars, sensationalises Gambier's alleged neglect of duty by showing the admiral as a religious zealot, preoccupied by personal worship at the cost of his earthly responsibilities. Cochrane enters the cabin with the ship's bemused boatswain, fresh from the fight, and exclaims 'D—n their Eyes they'l escape if we don't make haste'. Gambier, too engrossed to notice the interruption, reads an obscure passage from Sternhold and Hopkins's sixteenth-century edition of the Book of Psalms as a minister beside him raises his hands skyward and bemoans the 'wicked Dog' (i.e. Cochrane) who has disturbed their private study. Gambier's table renders one of the ship's guns useless and a map of the Holy Land replaces the usual navigational charts on the cabin wall, as other signs of neglect litter the floor: a telescope, the torn pages of the ship's log book, and a bundle of disused 'Congreve Rocketts'.

During the first decade of the nineteenth century a group of influential evangelical officers known as the Blue Lights gained prominence within the navy, promoting moral probity and regular religious observance across the ranks. At the same time, religious societies such as the Society for Promoting Christian Knowledge, the Naval and Military Bible Society and the Religious Tract Society continued to gather momentum and counted several high-profile naval captains among their supporters, with Gambier the most notable. *Sternhold and Hopkins at Sea* narrates a relatively minor and largely forgotten episode of naval history, but in the summer of 1809 the prospect of another high-profile military trial would have suggested a receptive audience for Williams's print. Moreover, its publication implies an audience well-versed in the finer details of naval and religious affairs, and suggests the degree to which both remained subjects of wider interest and scrutiny.

A TRUE BRITISH·TAR.

"*Damn all Bond-Street-Sailors I say, a parcel of smell-smocks!*
they'd sooner creep into a Jordan than face the French!—dam-me!

James Gillray, A True British Tar
Hannah Humphrey, 28 May 1795.
Hand-coloured etching. NMM PAF3811

At first glance, the stout figure in Gillray's etching appears to be a humorous yet affectionate portrait of a typical British seaman. On further inspection, however, the exaggerated physiognomy and the passing reference in the inscription to a 'Jordan' clearly identify the sailor as Prince William Henry, Duke of Clarence, the future William IV. The Duke served as a professional naval officer with Nelson, and later became Lord High Admiral; 'Jordan' was slang for a chamber pot, but here also makes reference to the Duke's mistress, the actress Dorothea Jordan (an unfortunate pun that gave rise to far cruder satires than this). Gillray's swipe at the dissolute behaviour of 'Bond-Street-Sailors' can thus be read as a barbed criticism of the Duke's own decadent lifestyle. Any real scorn is so deeply embedded in visual and verbal innuendo that Gillray would have nothing to fear by turning his attention to such a high-profile royal. However, Gillray's message is clear: the officer-duke is only able to assume such a noble stance by evoking the unofficial dress code and heroic figure of the ordinary sailor – the 'True British Tar' of the print's title.

James Gillray

Nelson's Victory: – or – Good News operating upon Loyal Feelings

Hannah Humphrey,
3 October 1798.
Hand-coloured etching.
NMM PAF3865

Gillray's first response to the Battle of the Nile turns on a group of prominent Whig grandees, distraught by the news of Nelson's momentous victory. The Whig party, guided by the forceful personality of Charles James Fox, had been early sympathisers of the French Revolution and vocal opponents of the war against Napoleon. Although the Whigs' support for the Revolution was tempered by the extreme violence of the Terror of 1793–94 (causing many to reject its ideals altogether), Fox and his closest allies were unable to escape their reputation as Francophile revolutionaries, bent on aiding the republican cause. In Gillray's image, the Marquess of Lansdowne and Fox both wear revolutionary *bonnets rouge*, and a portrait of Napoleon Bonaparte hangs on the wall behind Sir Francis Burdett.

Notably, Nelson is absent in all but the title of the print, although his influence is already overwhelming. As the most conspicuous and successful exponent of the war against revolutionary France, Nelson's likeness was increasingly called upon by printmakers as a focus for anti-Whig propaganda. Gillray's schematic print presents eight leading lights of the opposition as a succession of physical and moral afflictions provoked by Nelson's victory at Aboukir Bay: from the blindness of Burdett and drunkenness of the Duke of Norfolk, to the melancholy of George Tierney, and finally suicide, enacted by a deluded Fox who hangs himself in the corner as a note issues from his hand declaring 'Farewell to the Whig Club'. Gillray's exaggerated message of opposition disloyalty was unrealistic, yet continued to gain popular currency.

A victory more glorious and more complete is not recorded in the annals for our navy.

London Chronicle, 2–4 October 1798,
describing the Battle of the Nile.

CHAPTER 2

Nelson

Having risen to prominence following his actions at the Battle of St Vincent in 1797, Horatio Nelson directed naval forces at three major battles, winning each decisively. In August 1798, following a long chase across the Mediterranean, Nelson's fleet overwhelmed a French force at Aboukir Bay off the north coast of Egypt. The British victory at the Battle of the Nile, as it became known, marked a turning point in the war, made a national hero of Nelson, and encouraged the mighty Austrian and Russian empires to enter the fray against Revolutionary France. Three years later, Nelson led the main attack that captured the Danish fleet at Copenhagen, forcing the surrender of the city and opening up the Baltic Sea. Finally, in 1805, at a time of heightened national anxiety as Britain faced the continued threat of invasion from across the Channel, Nelson once again became the nation's defender.

James Gillray
Extirpation of the Plagues of
Egypt; – Destruction of
Revolutionary Crocodiles; –
or – The British Hero
cleansing ye Mouth of ye Nile
Hannah Humphrey,
6 October 1798.
Hand-coloured etching.
NMM PAF3893

Gillray's alternative reaction to the Battle of Nile, published within days of the acerbic satire of *Nelson's Victory*, offered a forthright celebration of Nelson's role in the battle – his slender figure in contrast to the Herculean nature of his task and the greatness of his achievements. Britain's newly appointed national hero wades knee-deep across the shallows of Aboukir Bay rounding up the biblical 'Plagues of Egypt' of the print's title, represented as fearsome 'Revolutionary Crocodiles'. The tricoloured beasts are easily vanquished by

Nelson's club of 'British Oak' – a patriotic allusion to the valuable natural resources upon which the navy depended, and a recurring metaphor for the British fleet. Meanwhile, in the middle distance on the far right of the image, an inferno erupts from a crocodile's gaping jaws, which a contemporary, newspaper-reading audience would have recognised as the dramatic explosion of the French flagship *L'Orient*, an incident that subsequently became the focus of several paintings and prints of the battle.

The initial response of London's caricaturists was to portray Nelson as the conquering hero, while turning their satirical ire against Britain's enemies and politicians at home who had opposed the war. However, as public fascination with Nelson's character and moral conduct intensified, his personal life received greater attention. While other branches of visual and material culture continued to present an unambiguous image of Nelson the hero, the possibilities of caricature encouraged a more fluid, nuanced reflection on the man and his achievements. As Nelson's relationship with Emma Hamilton became widely known, satirists were quick to respond. Increasingly, Nelson became the punchline, as his vanity and sexual proclivities became subjects of fascination.

Following his victory and death at the Battle of Trafalgar, Nelson's reputation as a pompous and egotistical womaniser inevitably gave way to a more solemn reflection on the sacrifice he had made for his country. Overnight, he became, in Byron's words, 'Britannia's God of War'. The prints assembled here trace a now familiar trajectory of the celebrity life played out in the popular media, from initial adulation, to invasive scrutiny, to death, reflection and sanctity.

' My dress from head to foot is alla Nelson. Even my shawl is in Blue with gold anchors all over. My earrings are Nelson's anchors; in short, we are be-Nelsoned all over. '

Lady Emma Hamilton to Nelson in September 1798, shortly after they had met.[5]

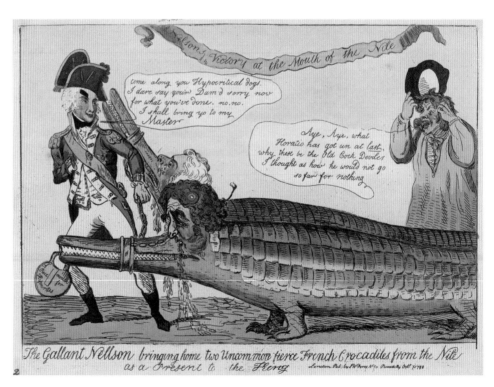

The Gallant Nellson bringing home two Uncommon fierce French Crocadiles from the Nile as a Present to the King
Samuel Fores,
7 October 1798.
Hand-coloured etching.
NMM PAF3889

In the months that followed the British victory at Aboukir Bay, Nelson continued to be defined by his Egyptian connection. Samuel Fores also adopted a crocodile theme to emphasise the exotic nature of Nelson's achievements at Aboukir Bay. In this print, the chained beasts of the Nile are given the heads of anti-war politicians Fox and Sheridan, whose 'crocodile tears' of remorse fail to impress Nelson. The 'Gallant' admiral is again shown as the servant of the pro-war Tories: 'Come along you Hypocritical dogs', he berates as he leads the disgraced pair in chains to George III, 'I dare say you'r Dam'd sorry now for what you've done.'

Dresses a la Nile respectfully dedicated to the Fashion Mongers of the day, William Holland, 24 October 1798.
Hand-coloured etching. NMM PAF3864

In the months after the battle, topical references to Egypt and the Nile were not limited to prints. Clothes, textiles and ceramics inspired by Nelson's Mediterranean exploits became favourite items among fashionable patriots. In an age where it was commonplace to display patriotic sentiment and political allegiance through dress and other accoutrements, society ladies (foremost among whom was Nelson's mistress, Lady Emma Hamilton) hurried to dress themselves 'a la Nile', with naval and Egyptian themed shawls, ribbons, hats and jewellery in honour of the hero of Aboukir Bay.

Dresses a la Nile, published anonymously towards the end of October 1798, gently mocks the Egyptomania of British 'Fashion Mongers'. Presented in the manner of contemporary fashion prints, it shows a young couple decked in the latest regalia. The woman, cocooned within a white crocodile pattern dress and with 'Nelson and Victory' labelled onto her headgear, carries a large muff with a small crocodile head representing the mouth of the Nile. Perhaps even more absurdly, her companion wears a crocodile skin suit, complete with frilly crocodile boots and a bright yellow crocodile balancing on his hat. Neither model appears entirely comfortable with their outfits, yet both appear determined to fulfil their fashionable duty. As a satire on late eighteenth-century sartorial excess, it is unsurpassed.

James Gillray
The Hero of the Nile

Hannah Humphrey, 1 December 1798.
Hand-coloured etching.
NMM PAF3888

When, in November 1798, Nelson was created Baron Nelson of the Nile and Burnham Thorpe and awarded a pension of £2000 a year, Gillray returned his attention to the Hero of the Nile, this time with a more ambivalent eye. Three months after the first appearance of *Extirpation of the Plagues of Egypt*, one of Gillray's most popular prints, the artist produced a full-length portrait print, presenting Nelson as a man tarnished by vanity and a disproportionate sense of his own importance. Nelson relished the adulation he received upon returning to London and appeared in full uniform at every opportunity. But rather than win further acclaim, Nelson's peacock-like behaviour often led to aggressive lampooning.

In Gillray's portrait, despite being on deck surrounded by gun smoke, the Admiral wears his dress uniform, an array of decorations and the regalia of the Order of the Bath. In contrast to the unqualified admiration of his earlier caricature, Gillray presents Nelson as a rather frail figure, bordering on the ridiculous as he is dwarfed by the material signs of his recent success, including the distinctive (and, in Gillray's hands, oversized) *chelengk* granted by the Sultan of Turkey, displayed proudly in his tricorne. Nelson's appearance is as preposterous, Gillray seems to suggest, as the two followers of fashion in *Dresses a la Nile*. The coat of arms beneath the figure, flanked by Jack Tar and a British Lion, commemorates Nelson's generous pension with a bulging purse and Nelson's personal motto (here included with some irony): *Palmam Qui Meruit Ferat* ('Let he who merits the palm bear it').

After meeting in Naples in 1798, Nelson and Lady Emma Hamilton began a passionate and increasingly public relationship. Their affair was tolerated by Emma's much older husband, Sir William, to the extent that the three travelled across Europe together and lived in an amicable ménage. By the time they returned to England, in November 1800, Emma was heavily pregnant with Nelson's child. The couple's unusual domestic arrangement was a gift to caricaturists and satirical writers casting a watchful eye over Nelson's every move.

Isaac Cruikshank

A Mansion House Treat – or Smoking Attitudes

Samuel Fores,

18 November 1800.

Hand-coloured etching.

NMM PAF3887

Tobacco smoke and innuendo fill the air in Isaac Cruikshank's bawdy take on Nelson's unconventional private life, which presents the trio in company with Sir William Staines, the recently elected Lord Mayor of London, and the Prime Minister, William Pitt. As William Hamilton struggles to light his tobacco, a tar waiting on the group observes that his pipe is too short and 'quite worn out'. Meanwhile Emma, sitting with her back turned towards Pitt and her cuckolded husband, agrees. 'Tho the old man's pipe is allways out', she confesses to Nelson, 'yours burns with full vigour'. 'Yes, yes', her lover replies eagerly (whilst holding the most suggestive pipe of all), 'I'll give you such a smoke I'll pour a whole broadside into you'. Cruikshank leaves little to the imagination, and his willingness to lampoon two of the most celebrated figures in Europe in such a crude way is striking. However, the tenor of the artist's print is not wholly critical as, ultimately, his satire also helped to bolster Nelson's reputation as a man of 'full vigour', in contrast to the frail protagonist of Gillray's *Hero of the Nile* two years before. The title of Cruikshanks's print refers both to the Lord Mayor's official residence and to the often erotic classical character poses, or 'Attitudes', for which Emma Hamilton was renowned across Europe (one of which she adopts in this print).

'Ah where, & ah where, is my gallant Sailor gone?
'He's gone to Fight the Frenchmen, for George upon the Throne.' *DIDO, in Despair!* 'He's gone to fight ye Frenchmen, t'loose t'other Arm & Eye,
'And left me with the old Antiques, to lay me down, & Cry.'

James Gillray
Dido in Despair!
Hannah Humphrey,
6 February 1801.
Hand-coloured etching.
NMM PAF3874

Lady Hamilton in a classical pose, engraved by Friedrich
Rehberg after Tommaso Piroli, from *Drawings Faithfully
Copied from Nature at Naples*, London, 1794. Engraving.
NMM PAD3218

Emma Hamilton's 'Attitudes' also provide a central motif for
Gillray's comic and cruel satire *Dido in Despair!*, which sees
the scandalous pair cast as the ill-fated lovers of the *Aeneid*.
In Virgil's narrative, the Trojan hero Aeneas falls in love with
the Queen of Carthage (a lead role played here by Lady
Hamilton) before abandoning her in order to found Rome,
after which Dido kills herself with Aeneas' sword. Gillray
plays with heavy irony upon the parallels between Virgil's
poem and Nelson and Emma's very public relationship,
transforming a first-century BC epic into a modern farce.
Gillray pictures the moment when Nelson resumed his duties
in January 1801, leaving behind his distraught mistress.

Emma's ungainly appearance makes another ironic allusion
to the celebrated classical poses performed in her youth, a
book of which had proved enormously popular when
published by Friedrich Rehberg in 1794. Alluding to
Rehberg's popular prints, Gillray includes a volume of *Studies
in Academic Attitudes* open on the couch below the window,
showing a reclining woman quite unlike the graceless figure
gesticulating on the bed. Gillray's Emma cries out in despair
at the loss of her 'gallant Sailor': 'He's gone to fight ye
Frenchmen, t'loose t'other Arm & Eye,' she bemoans, 'And
left me with the old Antiques, to lay me down, & Cry.'

A keen connoisseur, William Hamilton had acquired one
of the finest collections of antiques. But as Gillray observes, a
fondness for antiquities could be an unfortunate hobby for
an aged man with a young wife. Various lewd and broken
figurines – travesties of works from Hamilton's collection –
lie neglected at Emma's feet as the oldest 'antique' of all lies
asleep in his bed, seemingly oblivious to the drama unfolding
around him.

Rarities from Abroad!!!
J Garbamati,
18 March 1801.
Hand-coloured etching.
NMM PAF3927

Rarities from Abroad!!! refers to Nelson's second major victory at the Battle of Copenhagen. In 1801, although only the second-in-command, Nelson's initiative in the heat of battle ensured the Danes were defeated. The Baltic Sea was opened to British trade, removing Denmark from a northern alliance that had embargoed all trade with Britain. Crowds cheer as a British naval officer, bearing a passing resemblance to Nelson, returns home with four animals in tow. As with other prints that reference the nations of northern Europe, each animal represents a defeated Baltic state: a collared Russian bear, a pair of cubs representing Sweden and Denmark, vassals of Russia, and a dejected double-headed eagle of Prussia. Welcoming the parade of national animals is the rustic figure of John Bull, accompanied by a patriotic Scotsman who plays 'The Embargo Lilt' on his pipes.

Foreign Amusements or the British Lion on the Watch
Samuel Fores,
1 January 1801.
NMM PAF3936

Britain's reliance on shipbuilding resources from the Baltic states ensured that the nations of Northern Europe had long been a subject of caricature. In 1800 the Russian emperor Paul I formed a League of Armed Neutrality with Prussia, Sweden and Denmark. The League posed an immediate threat to essential supplies of naval stores by declaring the Baltic Sea closed to British commerce. Observing the stand-off prompted by the League's declaration, *Foreign Amusements or the British Lion on the Watch* appeared in Samuel Fores's Piccadilly print shop within days. The print, simple in technique and direct in its message, represents the mighty Russia as Paulo the Bear gathering the ships of the Baltic nations under lock and key as a vigilant British lion looks on, ready to pounce. In response, the Admiralty sent a British fleet to the Baltic to tame the Russian bear and re-open the region to trade. The decisive British victory that ensued at the Battle of Copenhagen became Nelson's next great triumph.

the Death of ADMIRAL-LORD-NELSON – in the moment of Victory!

this Design for the Memorial intended by the City of London to commemorate the Glorious Death of the immortal Nelson, is with every sentiment of respect, humbly submitted to the Right honble the Lord Mayor & the Court of Aldermen.

James Gillray, The Death of Admiral-Lord-Nelson – in the moment of Victory!

Hannah Humphrey, 29 December 1805. Hand-coloured etching. NMM PAF3866

Gillray's brilliant response to the news of Nelson's death shows him fatally wounded, surrounded by the chaos of battle, as a winged figure heralds the British victory and signals Nelson's immortality. As he draws his final breath, Nelson is supported by a forlorn Captain Hardy (pointedly resembling George III) and by a grief-stricken personification of Britannia, whose features would have been immediately recognisable to those familiar with Gillray's modus operandi as those of Emma Hamilton.

The hubbub around the dying figure may be exaggerated to the point of absurdity, but the sense of personal and national grief conveyed by Gillray's print is no less authentic. The real target of the artist's satire, it seems, is not Nelson (who finds redemption in death), but the unbridled and, at times, unseemly race to institutionalise the loss of a national hero. An inscription beneath the image informs the viewer that the allegory is, in fact, a proposal for an official memorial for the City of London, intended 'to commemorate the Glorious Death of the immortal Nelson'. In the event, Gillray's overstated tribute proved prescient, as many such proposals came to light in the wake of Nelson's death and funeral.

William Woollett after Benjamin West
The Death of General Wolfe
1 January 1776. Hand-coloured engraving. NMM PAH7700

The Death of Admiral-Lord-Nelson reveals Gillray's profound understanding of the forms and conventions of Academic and modern history painting, as well as his conviction that the art of caricature could aspire to the same kind of intellectual and creative status as other, more elevated forms of art. More specifically, *The Death of Admiral-Lord-Nelson* parodies one of the most successful images of the late eighteenth century: Benjamin West's epic depiction of the death of General James Wolfe at the Battle of Quebec in 1759, which West painted in 1770, and which became established internationally as a model of heroic death when engraved by William Woollett. The continued familiarity of West's iconic image well into the nineteenth century allowed Gillray to lampoon the carefully staged drama of Wolfe's battlefield demise.

After seeing Wollett's print, Nelson is said to have asked West that, should he also die in battle, West would paint him in a similarly epic manner. In the event, West produced several paintings of the death of Nelson, each of which was engraved, including the largest and best-known, *The Death of Nelson* from 1806. Although serious in intent, West's image of Nelson bears similarity, in turn, to the composition and personnel of Gillray's earlier caricature.

James Heath after Benjamin West
The Death of Nelson
1811.
Engraving.
NMM PAH8031

George Woodward and Thomas Rowlandson
The Brave Tars of the Victory, and the Remains of the Lamented Nelson
Rudolph Ackermann,
9 December 1805.
Hand-coloured etching.
NMM PAF3756

William Holland
The Sailor's Monument to the Memory of Lord Nelson
Hand-coloured etching.
NMM PAG8562

The fierce loyalty that Nelson inspired among those who served under him was a recurring feature of his posthumous appearance in caricature. In *The Brave Tars of the Victory*, the result of a fruitful collaboration between George Woodward and Thomas Rowlandson, two tars deal with the death of their hero in a typically bluff and honest manner. The crew of the *Victory* had insisted on bringing Nelson home themselves, rather than move his body to a faster frigate. Jack, on the right, leans protectively over Nelson's coffin as he reassures his comrade that he will watch over his precious cargo until it arrives safely in England whereupon, he predicts, 'his monument will be erected in the heart of every Briton'. Woodward and Rowlandson's print appeared just days after Nelson's battered flagship finally reached home.

In *The Sailor's Monument*, a veteran of Trafalgar rejects the commercialism and pomp that surrounded Nelson's official monument at St Paul's by erecting his own garden tribute to his 'Noble Companion'. The absurdity of 'poor Jack's' emblematical tribute, topped by an 'Englishman's Heart', is at once comic and affectionate, acknowledging the devotion to Nelson shown by those who had served under him. Ever loyal (and in this case demonstrating an uncommon sensitivity) the sailor's personal commemoration also refers implicitly to the wider furore over the preservation and presentation of Nelson's legacy. By focusing on the no-nonsense actions of Nelson's men, sympathetic caricatures such as *The Brave Tars of the Victory* and *The Sailor's Monument* foreground the other icon around which patriotic fervour could coalesce: the ordinary British sailor.

242

An English JACK-TAR giving MONSIEUR a Drubbing.

Published Nov.r 11th 1788. by Rob.t Sayer, 53, Fleet Street, London.

An English Jack-Tar giving Monsieur a Drubbing
Robert Sayer, 11 November 1788. Mezzotint.
British Museum 2010,7081.1003

Here, Jack Tar appears in a 'droll', a form of humorous mezzotint that was popular in the eighteenth century, which shares many of its bawdy themes with the hand-coloured caricatures that went on to dominate the market. This relatively late example shows Jack thrashing a foppish French visitor at the door of a Portsmouth tavern, as Admiral Keppel looks on approvingly from the sign above. The mezzotint was first published anonymously in 1779, during Keppel's court martial after the Battle of Ushant.

Nine years later, Robert Sayer re-issued the plate, adding new signs above the tavern door advertising 'Rodney's Cordial' and 'Hood's Intire' – both allusions to the more recent British victory at the Battle of the Saintes, and to Admiral Hood's subsequent complaint that Rodney had not gone on to destroy the entire French fleet. Despite the various spats between admirals, this print implies that the burly tar remains a constant if brutish force.

CHAPTER 3

Jack Tar

The warts-and-all image of the ordinary sailor as an unpolished but reliable and fiercely loyal pillar of the navy emerged during the second half of the eighteenth century. Like his landlubber cousin John Bull, Jack Tar was a product both of a period of conspicuous military success and of a print culture that was ideally suited to the portrayal of national stereotypes; his rise to prominence signalled an end to the era in which Britain's naval heroes had, without exception, been officers. The greatest champion of Jack Tar and the principal beneficiary of his popularity in print was undoubtedly the London publisher Thomas Tegg, whose imprint appears repeatedly in the images that follow. The myriad caricatures of Jack that emanated from Tegg's Cheapside print shop traded on innuendo, nautical jargon and a belief in the virtuous simplicity of the sailor to create an affectionate everyman image of the British tar as a straight-talking and dependable warrior. The most celebrated incarnation of the loyal and fearsome sailor, Gillray's *Fighting for the Dunghill*, showed Jack beating a revolutionary Frenchman – and then Napoleon himself – from the face of the world (see page 3). However, in an era when regular naval success was accompanied by mutiny and the rise of radical politics at home and abroad, the tar's unruly nature was also a potential threat to the social order that had created him. The prospect of conflict between the ranks, and Jack's dissolute behaviour when ashore, challenged the patriotic myth of Britain's seamen.

Monsieur sneaking Gallantly into Brest's sculking Hole
W Richardson,
c1778.
Etching with engraving.
NMM PAD4791

This anonymous etching offers an optimistic view of British achievements at the Battle of Ushant in July 1778, which the inscription describes as a mere opening shot – a 'preliminary salutation' from Jack Tar to his old enemy. Subsequent British victories ensured that the image of the British tar, shown here wielding a cat-o'-nine-tails at a farting French sailor, would become far more prominent during the next three decades.

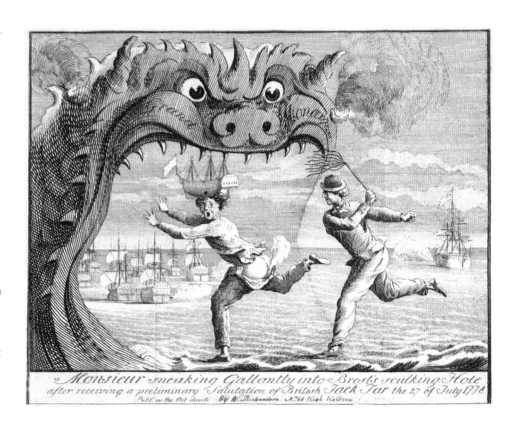

Monsieur sneaking Gallantly into Brest's sculking Hole after receiving a preliminary Salutation of British Jack Tar the 27 of July 1778.
Pub.d as the Act directs by W. Richardson N.º 68 High Holborn

Thomas Rowlandson
Nautical Politeness, or British
sailors perusing the
Dispatches from Cadiz
Samuel Fores,
c1808.
Hand-coloured etching.
NMM PAF3743

The mannered and overly polite exchange between Sir John
Moore, commander of the British army in the Iberian
peninsula, and Admiral Rosily following the surrender of a
French squadron at Cadiz, is rendered absurd when seen
through the eyes of three no-nonsense sailors below deck. The
officers' flowery rhetoric prompts one of the sailors to pen his
own reply to Rosily: 'Mounseer, I had the honor of your [sic]
this morning, and if you don't surrender by six in the evening,
I'll be d—d if I dont blow you up. Yours to command, Jack
Junk'. As well as lampooning the niceties of military etiquette,
recalling the dynamics of the much earlier print, *Count de
Grasse delivering his Sword* (see page 13), Rowlandson's
caricature also makes a more topical reference to the role of
the navy at the beginning of the Peninsular War, and to the
controversial Convention of Cintra, which allowed a defeated
French army free passage while the British fleet looked on.

Charles Williams
An Irish Pilot or Steering by
Chance
Thomas Tegg,
August 1812.
Hand-coloured etching.
NMM PAF3751

The British navy was made
up of a number of
nationalities, with sizeable
Irish, Welsh and Scottish
contingents, as well as
seamen from Scandinavia,
America and sometimes
even France. The diversity
of the lower-deck was well
known, and served as a
useful source of nationalistic
humour.

Charles Williams, Makeing a Compass at Sea, or the Use of a Scotch Louse
Thomas Tegg, c1812. Hand-coloured etching.
NMM PAG8606

Williams's *Makeing a Compass at Sea* draws its humour from Joe Miller's *Universal Jester*
(1810), a popular early nineteenth-century joke book that reveals a seemingly inexhaustible
appetite for hackneyed national and regional stereotypes. In this instance, it is the unflinching
national pride of the Scots that comes to the fore after the ship's compass has been broken.

**Thomas Rowlandson after
George Woodward**
The Welch Sailor's Mistake or
Tars in Conversation
Thomas Tegg,
30 June 1808.
Hand-coloured etching.
NMM PAF3757

'... And so then do you see
David, we sprung a leak' a
sailor declares at the end of
what we can presume was a
very long story, to which a
Welsh messmate replies:
'Cot pless us – and save us
– did you? and a ferry coot
fetchitable it is'. Wales'
choice of a vegetable as a
national symbol remains a
source of low-quality
humour.

George Woodward

An English Sailor at a French
Eating House

Samuel Fores,
30 May 1805.
Hand-coloured etching.
NMM PAF3842

**Thomas Rowlandson after
George Woodward**

The French Admiral on board
the Euryalus

Rudolph Ackermann,
11 December 1805.
Hand-coloured etching.
NMM PAF4006

For centuries, the British and French have defined one another according to their contrasting eating habits. These two related prints revolve around a comic play on words, and on the visual contrast between the unpolished but well-provisioned table of the British tar and the refined but insubstantial fare served up by his enemy. In the first, Jack Tar is confused by the indecipherable dish he has been given

in a French eating house. Meanwhile, on board the *Euryalus*, hours after the Battle of Trafalgar, three well-fed sailors sit at a makeshift table, drinking beer and feasting on a hearty dinner of British beef. A captured admiral, Pierre-Charles Villeneuve, is astonished at their ample provisions and now understands how his enemy were able to fight so well: 'No wonder,' he exclaims, 'you eat dam vel – and you drink dam vel!'

> My whack of prize money at a moderate calculation will be about fifty thousand pounds, which for a younger brother is not a bad fortune to have made.

Captain Charles Paget to his brother, 22 February 1805.[6]

Equity or a Sailors Prayer before Battle
Thomas Tegg,
c1805.
Hand-coloured etching.
NMM PAF3761

Equity or a Sailors Prayer before Battle offers a more challenging view of naval affairs by turning the fearless, straight-talking character of the British tar towards an adversarial and potentially threatening exchange with an officer on deck. Mistaking a sailor's prayer before battle as a sign of fear, the young lieutenant is soon corrected. 'Afraid!' cries the sailor. 'No! I was only praying that the enemys shot may be distributed in the same proportion as the prize money, the greatest part among the Officers.' The uneven distribution of prize money – the rewards distributed from the proceeds of a captured enemy ship – was a longstanding source of complaint. While the captain of a ship could expect to make a fortune through prizes, the non-officer crew received a tiny fraction of the whole, despite facing the same danger. Naval pay, including prize money, had been one of the complaints of the mutineers at Spithead and the Nore in 1797, the latter of which ended with ruthless reprisals.

No less contentious than the sailor's prayer is the response of another crew member: 'Why don't you sing Amen to that, Tom,' he whispers to a third seaman. The implied radicalisation of Jack Tar and his messmates makes *Equity or a Sailors Prayer before Battle* an unusually frank caricature of naval politics and policy. While challenging the more common portrayal of Jack as a cheerful simpleton, the print also appeals to the goodwill shown towards Britain's sailors immediately after Trafalgar by foregrounding their conspicuous courage in the face of enemy action.

The prevailing image of the British sailor ashore as either a bungling yokel of the sea, incapable of making any distance on land, or a drinking and whoring delinquent, can be understood as complementary aspects of the same conservative stereotype. While the recurring image of the clueless sailor on horseback reinforced the notion that the true British tar was only ever really at home aboard ship, the rude exploits of Jack about town (whose behaviour was sanctioned when confined to the impolite dockside areas of St Katherine's, Wapping and Portsmouth) may be viewed as a necessary expression of the strong constitution that was required for battle. The popular image of Jack Tar was undoubtedly informed by actual or reported behaviour, but above all else, served to reassure. The caricatures that follow, all of which post-date the serious challenge to naval discipline spearheaded by the Spithead and Nore mutinies of 1797, appear to celebrate the simple virtues of the ordinary sailor while distinguishing him, intellectually and socially, from the officers he served. At the same time, they forcefully characterise the British sailor as either too incompetent or too intoxicated by women and liquor to pose any real political threat.

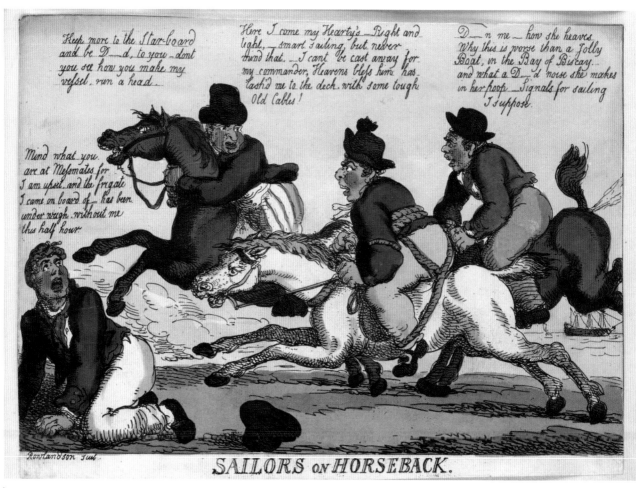

Thomas Rowlandson
Sailors on Horseback
Thomas Tegg,
1811.
Hand-coloured etching.
NMM PAG8619

The image of the bungling sailor on horseback was a recurring theme, and a favourite subject among Thomas Tegg's stable of caricaturists. As three startled horses gallop and fart uncontrollably along the seashore, their respective riders each describe their predicament in broad nautical English – none more eloquently than the third intrepid horseman: 'D—n me – how she heaves. Why this is worse than a Jolly Boat, in the Bay of Biscay. and what a D—d noise she makes in her poop – Signals for sailing I suppose.'

Sailors in a Calm

William McCleary, 1803.
Hand-coloured etching.
NMM PAF3806

Published by William McCleary from his premises at 32 Nassau Street, Dublin, *Sailors in a Calm* is a more or less faithful copy of a print published by Thomas Tegg in London earlier the same year, after a design by Woodward. McCleary's edition includes more text to extend the joke as three hapless tars struggle to spot the difference between their ship and the poor, overloaded 'crazy dutch built weather beaten old cast off Hulk' that they have managed to borrow from a local farmer.

Piercy Roberts after George Woodward

An Enquiry after Stretchit in Gloucestershire or the Sailors Reply

Piercy Roberts, c1803. Hand-coloured etching.
NMM PAF3859

This cheap print has been etched by Roberts with a coarseness that befits the crude humour of Woodward's design. Around the same time, their rival Charles Williams produced his own version of the same joke. There is no town of that name in Gloucestershire.

Though made as individual satires by different artists, the following few images suggest how caricatures on a common theme could be viewed together as part of a broader satirical narrative. In this instance, the story follows the hedonistic adventures of a sailor on shore leave.

William Elmes
Jack Jolly steering down
Wapping in Ballast trim
Thomas Tegg,
1813.
Hand-coloured etching.
NMM PAF3787
The night is young as a very jolly Jack Tar rides into Wapping with a woman on each arm and a large bottle of 'British spirits'.

JACK JOLLY steering down Wapping in Ballast trim.

Jack in a White Squall,
Amongst Breakers – on the
Lee Shore of St Catherine's
Thomas Tegg,
16 August 1811.
Hand-coloured etching.
NMM PAF3862

JACK in a White Squall. Amongst Breakers—on the Lee Shore of St. CATHERINE's

After a long day in the disreputable Thames-side area of St Katherine's, Jack finds himself in a squall: caught between an angry landlady who wants paying, and an irate prostitute who also has a score to settle. As the local nightwatchman approaches, Jack – fully spent, his pipe broken on the floor – prepares to ride the oncoming storm. Though crude in spirit, the composition of this and the previous image make a comic allusion to a far more elevated subject that was once popular among artists: the judgement of Hercules, in which the classical hero, approached by two contrasting goddesses, ponders a choice between a life of virtue and a life of pleasure. Casting Jack in the role of Hercules, though clearly ironic, serves as a timely reminder that the popular image of the British sailor was anything but one-dimensional.

Jack Tar

LAUNCHING A FRIGATE.

Thomas Rowlandson after Richard Newton
Launching a Frigate
Thomas Tegg,
1 February 1809.
Hand-coloured etching.
NMM PAF3786

In the distance, open arms and a roaring fire signal a warm welcome for a sailor returning home after months at sea, but the local procuress sends her best 'frigate' on a mission to capture the unwitting tar. The comic shift in style between the obese and disfigured brothel keeper and her genteel-looking employee is a subtle but highly effective device that distinguishes Rowlandson's work from the previous two images.

PORTSMOUTH POINT.

Thomas Rowlandson, Portsmouth Point
1811. Hand-coloured etching.
NMM PAF3841

Portsmouth Point exemplifies Rowlandson's unrivalled ability to extend caricature across a large and detailed scene. It looks back to the painted scenes of communal merrymaking by David Teniers and other Dutch artists of the seventeenth century, as well as to Hogarth's *March to Finchley*, in which a company of soldiers muster before marching out. Rowlandson's view unfolds, appropriately enough, between the tavern and the moneylender, where sailors of all ranks make the most of their final hours ashore. The figures become increasingly rowdy across the page, beginning with an officer saying farewell to his family on the right, and ending with a melee of carousing sailors and their 'wives' on the far left. The image represents an important moment for any ship, when the stores are replenished and the sails unfurled, as officers and ratings alike prepare to leave the pleasures of the land and return to sea.

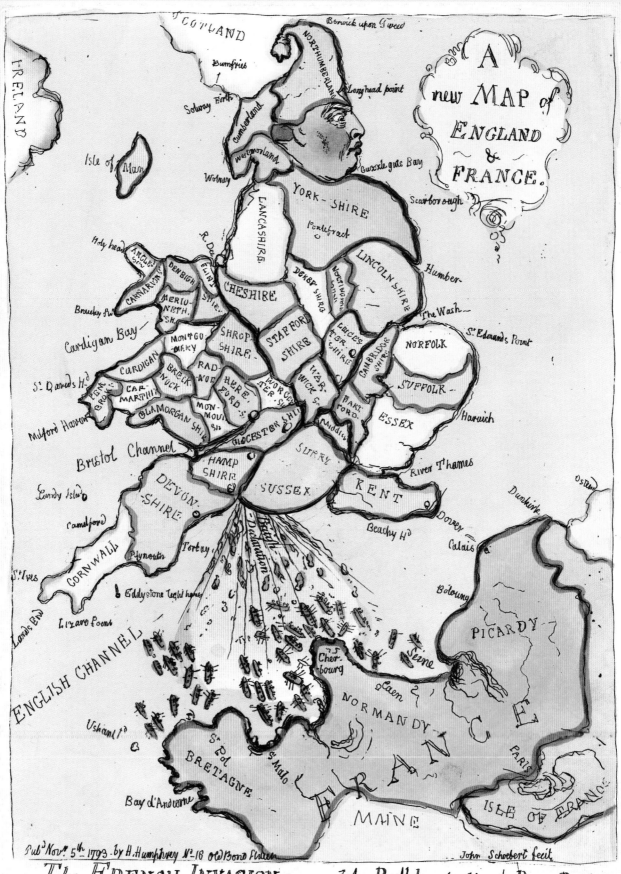

CHAPTER 4

Invasion

In one of the most memorable caricatures of the period, James Gillray presented his public with a partial map of the British Isles and north coast of France. To the English northeast (between 'Longhead Point' and 'Guzzleguts Bay') he added the distinctive profile of George III, whose portly features transform the nation, not only into a symbol of royal authority but into the irrepressible figure of John Bull, who squats over the north coast of France and repels the French fleet in heroic British style by issuing an unambiguous 'British Declaration' from the Portsmouth area – home of the British Channel Fleet. Urgent matters of national security are thus combined with cartography, royal portraiture, and an emerging pictorial tradition of John Bull as the crude but virtuous embodiment of the British people, to create one of the most enduring images in the history of Anglo-French relations.

The French Invasion; – or – John Bull, bombarding the Bum-Boats is one of the earliest artistic responses in Britain to the renewed prospect of invasion, following the outbreak of war with revolutionary France the previous February. The timing of Gillray's print engages with a longer national memory of invasion and defence. It was published on 5 November – an auspicious date that not only marked the anniversary of the discovery of the Gunpowder Plot in 1605, but also the day, in 1688, when William of Orange landed at Torbay with an army of 20,000 men during the so-called Glorious Revolution. William's arrival forced the departure of the Francophile James II, and set Britain against the Catholic powers of Europe, inaugurating a century of conflict and invasion scares that became especially pronounced during the wars against revolutionary France. John Bull's 'Declaration' against the French bum-boats is presented by Gillray as the latest historic British triumph against the threat of political and religious tyranny – a decisive act of defiance from the land of John Bull and George III against the revolution and regicide that was unfolding across the Channel.

James Gillray
The French Invasion; – or – John Bull, bombarding the Bum-Boats
Hannah Humphrey,
5 November 1793.
Hand-coloured etching.
British Museum Satires 8346

The practice of endowing a map with the physical characteristics of the region or nations represented was not new, but Gillray (working here under the alias 'John Schoebert') is more cavalier than most in his approach to cartography, rearranging the counties of England and Wales. Northumberland becomes a nightcap, Kent a boot, and Wales the tails of a splendid multicoloured coat. In an additional touch of cruel humour, the artist has represented France as the face of an emaciated old man, in contrast to the corpulent figure of John Bull who dumps wholeheartedly into the toothless mouth of his enemy.

A technically brilliant printmaker, Gillray here adopts a deliberately rough-edged style of etching as a way of underlining the coarse humour of his subject and the simple, patriotic virtue of John Bull. In this way, the print combines a visual and political immediacy with the kind of pictorial sophistication that few other artists could rival.

In the event, John Bull's bombardment was merely the opening salvo of a much longer campaign – the first of many caricatures on the theme of invasion to appear as the war with France continued into the next century. While some prints imagined the effects of a successful French landing (at once disastrous and comic), others focused on the scale of France's military ambition, highlighted the vulnerability of an unguarded coast, and foregrounded the role of the Royal Navy in the defence of the realm. The logistical and geographical challenges of mounting a full-scale invasion ultimately thwarted French ambitions, but a succession of smaller incursions, regarded as rehearsals for the main event, meant that the threat was never far from view.

In December 1796 a French fleet of transports carrying 15,000 troops was intercepted after being stranded in Bantry Bay, Ireland. One ship, the *Droits de l'Homme,* escaped but was later driven ashore on the Brittany coast. Harsh winds and blizzards took care of the rest and the would-be invasion ended in disarray. A few weeks later, in February 1797, Britain's coastal defences were put to the test again when French frigates landed over 1000 troops on the Welsh coast at Fishguard. The small and ill-disciplined invading army was quickly rounded up by Welsh militia. However, the potential consequences of another more successful attempt were felt across Britain, prompting a run on the banks and deeper economic repercussions.

The prospect of a full-scale invasion peaked on two occasions over the next decade: in 1797, as the French Directory turned their gaze towards Britain following a succession of victories on the Continent; and again from 1803, after the failure of the short-lived Treaty of Amiens, when Napoleon assembled a vast army and transport craft on the north coast of France. The idea of a full-scale invasion was often greeted with

Chart of La Manche by Nicolas Sanson
published with Alexis Hubert Jaillot's *Le Neptune François,* Paris, 1692.
NMM G223:1/22

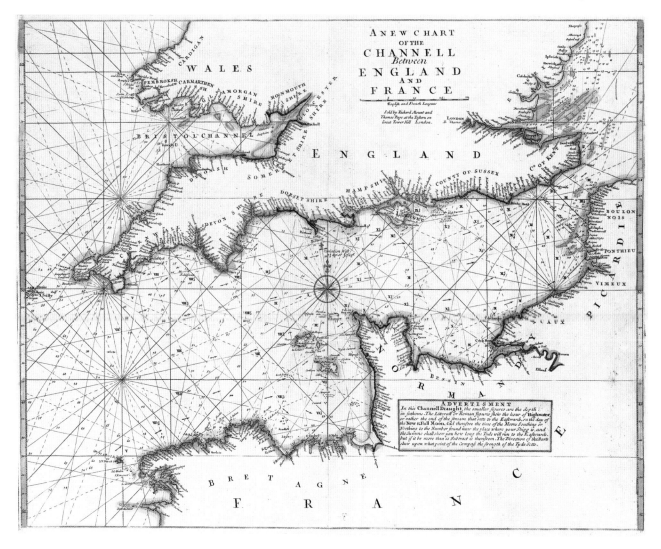

ridicule, or with the same crude complacency suggested by *John Bull, bombarding the Bum-Boats*, but even the most outlandish reports from across the Channel exploited a genuine and well-founded fear of attack. Faced with the growth of popular radicalism at home and rebellion in Ireland, the government were determined to harness the patriotic fervour that the threat of invasion (real or imagined) could inspire. The prospect of invasion from across the Channel and rebellion from within provided William Pitt's government with a rationale for far-reaching political intervention, including the suspension of habeas corpus in 1794 (and again in 1798), the introduction of Britain's first income tax, and the expansion of a disciplined county militia. During the summer of 1798 militia numbers in Britain more than doubled to 116,000; five years later, around half a million volunteers were armed and ready to defend the English coast.

James Gillray
End of the Irish Invasion; – or – The Destruction of The French Armada
Hannah Humphrey,
20 January 1797.
Hand-coloured etching.
British Museum Satires 8979

Inspired by the abandoned French attack at Bantry Bay in 1796, and by the powerful imagery of shipwreck that had been introduced into Britain from the Netherlands during the previous century, *End of the Irish Invasion* represents Charles James Fox as the carved figurehead of a dismasted ship, *Le Révolutionaire*, overcome by the patriotic blows of Prime Minister William Pitt (whose distinctive profile is silhouetted against the sun's fiery rays on the left) and other cabinet members. Fox, the hard-living Whig, was one of the most prominent and outspoken politicians of his age. His continued sympathy for the French Revolution

eventually forced him into retirement and prompted his opponents to portray him as a treasonous figure, bent on bringing the violence of the French Terror to Britain. As well as his confrontational politics, Fox's unkempt, swarthy appearance and dissolute behaviour made him a favourite target for caricaturists, thanks to whom he remains one of the most recognisable figures in British political history. His appearance here as the carved 'figurehead' of a dangerous faction within British politics is especially potent, and typical of a distinctive form of Tory patriotism refined by Gillray.

Towards the end of 1797, rumours of an imminent invasion began to gather momentum. Sightings of a French flotilla assembling off the Brittany coast were exaggerated en route to the London printing presses, and reports soon circulated of a great 'floating machine' under construction off the north coast of France, a quarter of a mile long and capable of carrying an entire army. Within days of the first news reports, rival printmakers produced a succession of fantastical interpretations of the fabled invasion machines that loomed unseen beyond the horizon. One printmaker, adapting an existing image, even interpreted the legendary French floating machine as a giant hot-air balloon – a citadel in the sky, complete with its own apartments, hospital and coffee house.

Robert Dighton

An Accurate Representation of the Floating Machine Invented by the French for Invading England
Robert Dighton, c1798. Hand-coloured etching.
NMM PAH7433

Robert Dighton's vision of a floating citadel, heavily armed and powered by an absurd combination of windmill and paddle wheel, is at once preposterous and powerfully direct in conveying the imagined, unquantifiable threat from across the Channel. Dighton plays on the same eyewitness claims for authenticity adopted by newspaper reports, citing as his source a drawing by a mysterious Monsieur Freville, 'just arriv'd' from France. The mock realism fostered by such prints is underlined by a handwritten note at the bottom of the page, probably added by Dighton himself, correcting the original inscription by claiming that the French had, in fact, contrived ten such vessels, each capable of carrying six thousand men.

> Various are the accounts we hear of their Floating Machines, it is even doubted if they are actually in existence, tho' it is beyond a doubt that they intend to attempt an invasion.
>
> Martha Saumarez writing to her husband, Admiral James Saumarez.7

> A Master of a neutral vessel arrived at one of the out-ports, declares he was in Saint Maloes eight days ago, and that a Raft, one quarter of a mile long, proportioned breadth, and seven balks deep, mounting a citadel in the centre, covered with hides, was nearly finished; and that a second upon a much larger scale, being near three quarters of a mile long, was constructing with unremitting activity.

The Star, 10 February 1798.

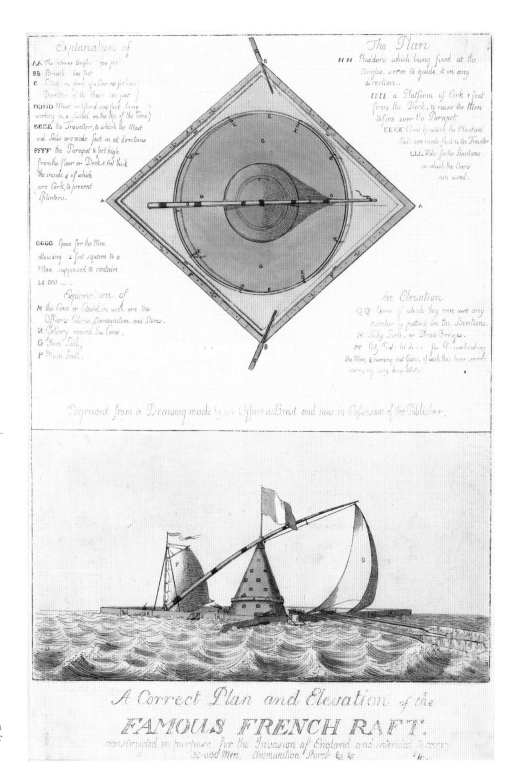

A Correct Plan and Elevation of the Famous French Raft. constructed on purpose for the Invasion of England
Samuel Fores,
1 February 1798.
Hand-coloured etching.
NMM PAD4060

This geometric raft is accompanied by the same eyewitness claims for authenticity – this time citing a drawing by an officer at Brest. The alternative views, elaborate key and detailed measurements are intended to convey an intimate knowledge of the craft, akin to the technical precision of a shipwright's plan.

The Storm rising : — or — the Republican FLOT

THE RAFT IN DANGER or the REPUBLICAN CREW DISAPPOINTED.

Isaac Cruikshank, The Raft in Danger or the Republican Crew Disappointed

Samuel Fores, 28 January 1798. Hand-coloured etching. British Museum Satires 9160

Invasion

James Gillray
The Storm rising; – or – the
Republican Flotilla in danger
Hannah Humphrey,
1 February 1798.
Hand-coloured etching.
NMM PAF3949

Once established in the public imagination, the great French 'Floating Machine' soon became part of the broader repertoire of political caricature, as a comic yet highly charged symbol of the revolutionary threat to the British Isles. Here Gillray incorporates the raft, complete with windmill propulsion and castellated defences, into a larger satire on the domestic political struggle between the government of William Pitt and the prominent radical Whigs led by Charles James Fox and Richard Sheridan. Fox, shown with his allies at a capstan, helping to draw the French invasion closer to the land, has even removed his coat so he can work with more gusto, revealing a French tricolour bow. Meanwhile, recalling the previous year's *End of the Irish Invasion*, Pitt forces the flotilla back by conjuring a storm with a patriotic offshore wind, and lightning charged with the names of the Royal Navy's admirals. Gillray's print appeared within days of an almost identical image by Isaac Cruikshank. By reversing the composition, turning Fox away from the viewer, Gillray seems to challenge his great rival in caricature by demonstrating the superior economy of his art – representing the leader of the opposition with little more than an eyebrow.

The twin rocks of Saint-Marcouf off the Normandy coast had been occupied by the Royal Navy as an anchorage for the blockade fleet and a forward base launching sorties to the mainland since 1795. In March 1796, the original occupying force, formed from the crews of two Royal Navy ships *Badger* and *Sandfly*, was joined by a Company of Invalids from Guernsey, comprising around fifty marines otherwise deemed unfit for service at sea. In the early hours of 7 May 1798, the tiny rocky islands were attacked by a heavy flotilla of French gunboats. By taking advantage of the unusual geography of the islands, the occupying British marines easily repelled the attack, imposing considerable losses on the enemy. Rowlandson's response appeared just days after the first reports of the action reached London.

In 1801 the new French leader Napoleon began to assemble an invasion force within sight of the English coast at Boulogne. The following year, the Treaty of Amiens brought a temporary ceasefire between Britain and France, but the possibility of invasion re-emerged with the recommencement of war in May 1803. The response of

Thomas Rowlandson
Rehearsal of a French Invasion as Performed before the Invalids at the Island's of St Marcou
Rudolph Ackermann,
18 May 1798.
Hand-coloured etching with aquatint.
NMM PAG8992

Rowlandson clearly delights in the news that the best efforts of the French military machine had been thwarted by a makeshift garrison and a vastly outnumbered company of invalids. For all its patriotic humour, however, he does not shy away from the reported violence of the action. Against the backdrop of celebrating peg-legged marines is the gruesome spectacle of French soldiers drowning by the dozen within sight of the French coast. The 'rehearsal' of

the print's title refers to Napoleon's planned full invasion of England – an event, Rowlandson implies, that would provoke a similar outcome. The artist had addressed the theme of invasion in an earlier companion print to the *Rehearsal*, titled *England invaded, or Frenchmen naturalized*, in which a landing force of French soldiers are 'naturalized' by being run through by an outnumbered but indomitable English militia.

> The carnage of the enemy was very considerable; we have accounts from the coast this morning, which states their loss to have been 1,200 men. The decks and boards that have floated from the sunken vessels are dyed with the blood of the sufferers.

London Packet or New Lloyd's Evening Post, 11 May 1798.

Invasion

James Gillray after John Dalrymple
Consequences of a Successful French invasion, No.1–Plate 2d. – We explain de Rights of Man to de Noblesse
1 March 1798.
Hand-coloured etching.
NMM PAG8509

The prospect of a successful invasion provided many opportunities for artists to contrast the perceived values of the British way of life with the apparent iconoclasm and violence of revolutionary France. In this, the second of four related scenes on the bloody consequences of a French occupation, based on a description provided by Sir John Dalrymple, Gillray imagines the destruction of the British parliamentary system, as marauding French soldiers pillage the House of Lords (the first in the series shows a comparable scene in the Commons). The throne has been replaced by a guillotine; the mace with the busts of anti-royalist heroes of French history. Such a disaster would signal the failure of the Royal Navy – a terrible notion alluded to in Gillray's print as the seventeenth-century Armada tapestries that adorned the House of Lords are torn down by French sailors and soldiers under the orders of a French admiral who exclaims 'Me like not de Omen; destroy it'.

Britain's caricaturists was markedly different from the anti-invasion propaganda of the 1790s, reflecting the different nature of the threat and the political and cultural ascent of Napoleon Bonaparte, who now took centre stage.

Napoleon committed huge resources to the planned invasion, including the creation of a new basin at Boulogne, resolute in his belief that only a few miles of water lay in the way of total French victory. Within a few months, a formidable French landing force of 80,000 men had been assembled along the French coast, and a vast flotilla of boats and barges constructed to transport the troops across the Channel. By the spring of 1805, Napoleon had assembled a mighty invasion force of more than 160,000 men, and medals were struck in anticipation of an imminent French victory.

Invasion fears reached their zenith in 1803, prompting Fox to remark of the hysteria that 'a picture of a People so terrified as we have been was never before exhibited'.[8] The British government responded by reinforcing the militias, as much to allay public fears as for strategic advantage. The navy followed suit, under Nelson's popular command, with a series of attacks on French ports which, although indecisive, were sufficient to persuade Napoleon to postpone his invasion plans. In the event, the few miles of water between the two nations proved more of a barrier than the emperor had imagined, but only a decisive naval victory could finally quash the threat of invasion. The Battle of Trafalgar in October 1805 destroyed French sea power and immediately signalled the end of Napoleon's English adventure.

> Let us be masters of the Straits for six hours
> and we shall be masters of the world.

Napoleon to Admiral Latouche Tréville in 1804.[9]

> I do not say that the French cannot come,
> I say only they will not come by sea.

Attributed to Admiral St Vincent,
addressing the House of Lords in 1803.[10]

me will make dat jean Bull Tremble now I have found out de Grande Conveyance

Invasion

MY ASS in a BANDBOX

Robert Holborn
My Ass in a Band Box
Piercy Roberts,
c1803.
Hand-coloured etching.
British Museum Satires 10101

With the return of war and full mobilisation of the Royal Navy, fears of a colossal invasion force over the horizon became a distant memory, and Napoleon's continuing designs on England were increasingly represented as a laughable folly. Robert Holborn's *My Ass in a Band Box* is fairly typical of the caricaturists' response, and a good example of early nineteenth-century caricature at its most economical, where every mark on the page contributes to the

comedic and patriotic value of the image. It represents Napoleon as a solitary figure, oblivious to the absurdity of his enterprise as he sits motionless on the 'Ass' of France. The oversized hat and feather, the diminutive figure of Little Boney himself, and the blinkered and branded creature on which he plans to cross the channel – even the box in which he is stranded upon a calm sea – all help to characterise Napoleon as a farcical, self-defeating character.

James Gillray

Physical Aid, – or – Britannia recover'd from a Trance: – also – the Patriotic Courage of Sherry Andrew; & a peep thro' the Fog

Hannah Humphrey, 14 March 1803.

Hand-coloured etching.

NMM PAF3967

In 1801, Henry Addington became Prime Minster and leader of the government that would negotiate the Peace of Amiens – an arrangement that both sides understood to be a temporary respite in the war. Addington maintained a large peacetime army and navy, justifying both as necessary precautionary activities. In March 1803, relations between the two nations deteriorated, and rumours of a French expedition began to spread. Many saw Britain as the most likely destination for Napoleon's forces. Fox complained that the House had been deliberately left in 'utter darkness' and feared a return to war. His old ally Sheridan gave a rousing speech praising the British character, which was later published, followed by an *Address to the People*. Sheridan's bombastic *Address* reprinted a pivotal speech from his popular tragedy *Pizarro*, to contrast the motives of the French and British nations: 'THEY, by a strange Frenzy driven, fight for Power, for Plunder, and extended Rule— WE, for our Country, our Altars, and our Homes. THEY follow an ADVENTURER, whom they fear, and obey a Power which they *hate*—WE serve a *Monarch* whom we love, a God whom we adore.'[11]

Physical Aid plays on the complex response of parliament to the renewed prospect of war. As Britannia swoons,

overcome by the sight of an impending invasion by the French 'Buggabo's', Addington and Hawkesbury come to her aid as Sheridan (dressed as a Harlequin) stands ready to defend her from an approaching French army, as more soldiers gather on the opposite coast. Gillray ridicules Addington's claim that the French had no intention of invading Britain. Holding a bottle of gunpowder to Britannia's nose, the Prime Minister looks with concern at the nearing flotilla: 'Do not be alarm'd my dear Lady,' he says with faint reassurance, 'the Buggabo's (the Honest Genlemen [*sic*], I mean) are avowedly directed to Colonial service, – they can have nothing to do Here'. Fox (by now a marginal political figure) appears in the shadows, hiding behind his hat, unaware of the reasons for Britannia's faint: 'Dear me – what can be the reason of the Old Lady being awak'd in such a Fright? – I declare I can't see any thing of the Buggabo's!' Gillray also scorns Sheridan's patriotic bluster. Wearing a fool's cap, he shouts in defiance at the approaching army: 'Let 'em come!' he cries, a copy of the previous year's Treaty of Amiens discarded at his feet, 'single handed I'd beat forty of 'em!!!'. Finally, *Physical Aid* also sees an early appearance of Napoleon as Little Boney – the pocket-sized emperor, dwarfed in equal measure by his uniform and ambition.

Theatre Royal, England.

In Rehearsal, and meant to be speedily *attempted*,

A FARCE
In One Act, called THE

Invasion of England.

Principal Buffo, Mr. BUONAPARTE

Being his FIRST (and most likely his Last) Appearance on this Stage.

ANTICIPATED CRITIQUE.

THE Structure of this Farce is very *loose*, and there is a *moral* and radical Defect in the Composition. It boasts however confiderable Novelty, for the Characters are ALL MAD. It is probable that it will not be played in the COUNTRY, but will certainly never be *acted* in TOWN; where ever it may be reprefented, we will do it the Juftice to fay, it will be received with *loud* and *reiterated* burfts of----CANNON!!! but we will venture to affirm, will never have the Succefs of

JOHN BULL.

It is however likely that the Piece may yet be put off on account of the INDISPOSITION of the PRINCIPAL PERFORMER, Mr. BUONAPARTE. We don't know exactly what this Gentleman's Merits may be on the Tragic Boards of France, but he will never fucceed here; his Figure is very Diminutive, he Struts a great deal, feems to have no Conception of his *Character*, and treads the Stage very badly; notwithstanding which Defects, we think if he comes here, he will get an ENGAGEMENT, though it is probable that he will fhortly after be reduced to the Situation of a SCENE-SHIFTER.

As for the Farce, we recommend the Whole to be Cut down, as it is the Opinion of all good Political Critics, that it will certainly be

DAMN'D.

"Vivant Rex & Regina!"

London: Printed for J. ASPERNE, Succeffor to Mr. SEWELL, at the Bible, Crown, and Conftitution, No. 32, Cornhill, by E. MACLEISH, 2, Bow-ftreet, Covent-Garden

Price Two-pence; or 12s. the 100; or 1s. 6d. per Dozen.

Where may be alfo had, a Collection of all the Loyal Papers that have been Publifhed.

* * Noblemen, Magiftrates, and Gentlemen, would do well, by ordering a few Dozen of the above Tracts of their different Bookfellers, and caufe them to be ftuck up in the refpective Villages where they refide, that the Inhabitants may be convinced of the CRUELTY of the CORSICAN USURPER.

A Farce … The Invasion of England

Mock playbill printed for J Asperne by E MacLeish in Bow Street, c1803.
Hand-coloured etching.
NMM PBF5077

The unnamed author of a mock playbill advertising a new one-act farce at the Theatre Royal encapsulated the new feeling towards Napoleon brilliantly, casting the emperor as a delusional character on the stage of European politics, better suited to shifting scenery than playing the lead role. The bold typography looks convincingly like an advertisement for a genuine production currently 'in rehearsal', but the extent of the satire soon becomes apparent as Napoleon is compared to another, more successful star of the stage, John Bull. 'We don't know exactly what this Gentleman's Merits may be on the Tragic Boards of France,' declares one damning critic, 'but he will never succeed here'.

The COFFIN EXPEDITION or Boney's Invincible Armada Half Seas Over

Charles Williams

The Coffin Expedition or Boney's Invincible Armada Half Seas Over
Samuel Fores,
6 January 1804.
Hand-coloured etching.
NMM PAD4783

The quixotic nature of Napoleon's designs take a sinister turn in *The Coffin Expedition*, which combines the effortless destruction of French gunboats at Saint-Marcouf with the hopeless folly of Napoleon's mode of transport in *My Ass in a Band Box*. Boney's 'Grande Conveyance' has been transformed into a fleet of overloaded coffins sinking under their own weight and manned by a hapless invasion force dressed for their own death. Two British ships approaching from the horizon have little to do but observe the self-destruction of the French gunboats and comment on 'Boney's Crest, a Skull without Brains' on the masthead of each landing craft. The despairing claim of one doomed Frenchman, that the disaster had been engineered on purpose by Napoleon was often repeated. It is a grisly and comic caricature informed by the events and visual history of the previous decade.

Charles Williams
Jack Tars conversing with
Boney on the Blockade of
Old England
Walker,
December 1806.
Hand-coloured etching.
British Museum Satires 10623

Jack Tars conversing with Boney on the BLOCKADE of OLD ENGLAND.

In November 1806, his plans for invasion scuppered, Napoleon declared the whole of Britain captive and prohibited all cross-Channel trade with any part of his European empire in an attempt to ruin the British economy. Britain responded by blockading the entire French-controlled Continent and declaring all ships coming from French-controlled ports contraband. The result was two warring nations, each claiming to blockade the other – one by land, the other by sea. *Jack Tars conversing with Boney on the Blockade of Old England* exposes the futility of Napoleon's attempted embargo. As well as continuing a perennial cross-Channel quarrel, Charles Williams's satire contrasts the effectiveness of the Royal Navy's blockade by sea with

Napoleon's failed navy and futile attempt to curtail British trade by land. As the emperor declares his policy a success, two British seamen, incredulous at Boney's claim and well-supplied with rum and tobacco, reply: 'Why. what do you mean by that you whipper snapper – here's Tom pipes and I in this little cock boat, will Blockade you so that you dare not bring out a single Vessel; – Blockade indeed! you are a pretty fellow to talk of Blo[c]kading!' In the distance, John Bull applauds the heroic efforts of the British navy from the English shore: 'I cannot help laughing at the whimsical conceit' he shouts. Neither, it seems, could Williams's appreciative London audience.

Charles Williams after George Woodward
The Continental Dockyard
Thomas Tegg,
27 November 1807.
Hand-coloured etching.
NMM PAF3999

A few months later George Woodward addressed a similar theme with *The Continental Dockyard*, in which an enormous British ship of the line, bedecked with assorted trophies from the defeat of the Spanish Armada to Trafalgar, is contrasted with a decrepit French fleet, massed in disrepair on land.

THE CONTINENTAL DOCKYARD

Isaac Cruikshank

The Empress's wish or Boney Puzzled!!

J Johnstone,

9 August 1810.

Hand-coloured etching.

NMM PAD4784

France's impotence at sea became a recurring theme of the Napoleonic Wars and a favourite theme for many caricaturists. As Britain enjoyed dominion of the seas after the Battle of Trafalgar, Napoleon continued to build up the French fleet, but his ships were easily contained by a well-organised and motivated British blockade. Cruikshank's satirical view of Napoleon's waning power finds Boney unable to fulfil the desires of his empress, who grips an extended telescope and asks Boney for the 'little ship' out at sea. The object of her desire is a British patrol on a blockading station – visible, yet beyond the reach of her emasculated, posturing emperor. The lewd sexual undertones of Cruikshank's print mercilessly reinforce the notion that even a single British ship, apparently within range of Napoleon's firepower, was untouchable while the French fleet was locked at its moorings.

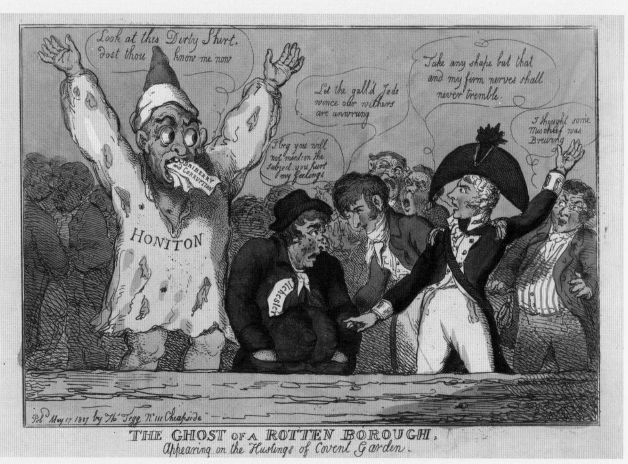

The Ghost of a Rotten Borough. Appearing on the Hustings of Covent Garden
Thomas Tegg, 17 May 1807.
Hand-coloured etching.
NMM PAF3996

In 1806 Thomas Cochrane, a successful naval officer and a man of radical political persuasion, contested the parliamentary seat of Honiton in Devon, a noted 'rotten borough'. Despite a population in the tens of thousands, only a few hundred individuals were able to vote, making it relatively easy for the local elite to influence the outcome of any election. Cochrane bemoaned high taxes and corrupt electoral practices, and refused to use bribes to win the seat. Unsurprisingly he lost by 259 votes to 124. Cochrane responded by sending a town crier through Honiton with the news that anyone who voted for him at the next election would receive ten guineas in return. The collapse of the 'Ministry of all the Talents' six months later instigated a general election, and a resounding victory for Cochrane at the second time of asking. However, when voters approached him, now safely in seat, he told them to expect 'not a farthing'.

In May 1807, the following year, Cochrane stood for election as MP for the seat of Westminster, seen as the most important constituency in the country due to its broad franchise and proximity to the capital. His opponents were Sheridan, a leading Whig; Elliot, the Tory candidate; and Burdett and Paul, two radical rivals. *The Ghost of a Rotten Borough*, produced a few days before the final poll, shows Cochrane resplendent in naval uniform, standing in front of his rivals. A ghost representing Honiton appears, reminding viewers of Cochrane's dubious 'radical' credentials. In its mouth are tickets referring to the practices that secured victory during his previous election.

The print reprises Macbeth's visitation by the ghost of Banquo, and the guilt of a man confronted with an unwelcome past through a supernatural figure. Invoking Shakespeare's protagonist, Cochrane exclaims: 'Take any shape but that, and my firm nerves Shall never tremble'.[12] Macbeth's words were especially resonant in the light of Cochrane's conduct during hustings at Covent Garden, where he wrested maximum advantage out of his naval reputation for bravery, boldness and firmness of action. Attended by large numbers of naval officers, Cochrane's speeches deliberately played up to a military ideal. He was critical of the Admiralty and abuses in naval administration, and cultivated an image as a warrior against corruption. The *Naval Chronicle* reported that he 'pledged himself to hunt down plunder, peculation, sinecure placemen, and pensioners, wherever he could find them'.[13] Cochrane's naval background allowed him to appear a patriotic hero – something his rivals could not hope to emulate.

Naval personnel held considerable sway in domestic politics. By alluding to Cochrane's 'firm nerves', this intricate print indicates how, as a candidate, Cochrane had attempted to distinguish himself as decisive and heroic, while also acknowledging that those credentials were fundamentally undermined by his prior electoral conduct. It also demonstrates how such prints did not always represent the majority view: one week after its publication Cochrane was elected to the second of the two available seats, alongside Burdett.

CHAPTER 5

Politics and the Navy

The navy was understood in parliament to be central to British interests, yet proved a politically contentious institution. State spending on the navy constituted as much as a quarter of the annual budget, and successive governments were charged with inefficiency and corruption by opposition politicians and radicals such as William Cobbett. Abuses of patronage and preferment were picked up quickly by satirical artists eager to forward political objectives. At the same time, conflicting professional and political interests of leading naval officers provoked caricaturists' ire. An officer's prospect of promotion was enhanced as a Member of Parliament, while naval officers were particularly attractive candidates for an electorate awash with naval patriotism. Between 1790 and 1820 more than one hundred MPs were naval officers, although nearly all saw their legislative responsibilities as subservient to their naval careers. Most continued to serve at sea, attending Parliament when possible, and supported the government of the day, taking care not to damage their prospects of promotion. However, a number were committed to Whig and more radical factions, taking centre stage in parliamentary debates and becoming outspoken figures of political opposition.

Naval Triumph or Favors Confer'd

J Harris,
13 November 1780.
Hand-coloured etching.
NMM PAH3331

The fortunes of Sir Hugh Palliser, rewarded despite his failure in battle, are contrasted with the neglect of British seamen, crippled and destitute in the service of their country. A noted Tory politician as well as an admiral, he was blamed in many opposition newspapers for the indecisive Battle of Ushant in 1778, as part of a broader media campaign against governmental incompetence. Although public hostility forced him to resign as an MP,

Palliser was subsequently appointed to the prestigious post of Governor of Greenwich Hospital, largely through the patronage of the Earl of Sandwich, the Tory First Lord of the Admiralty. In this dark and cutting satire, Sandwich, riding on the back of a ghoulish invalid, leads the richly attired Palliser by the hand towards the west gate of Greenwich Hospital, as various peg-legged seamen look on with disdain.

The most controversial aspect of the government's involvement in naval matters was impressment. During the eighteenth century the navy struggled to man its ships. While in peacetime the navy could rely on a relatively small number of well-trained seamen, in times of war the demand for manpower rocketed. Volunteers always provided the majority of the navy's personnel, but regular shortfalls forced the government to resort to the press gang. While some saw the policy as a necessary evil, its violence and abuses were vehemently attacked in pamphlets, newsprint and caricature. Although landsmen were sometimes taken by mistake, press gangs concentrated their efforts on finding experienced seamen, who were often taken from merchant ships. Nonetheless, the image of men being plucked from their homes or pulled off the street in order to fulfil the insatiable needs of the navy attained a powerful, mythical status in the popular imagination.

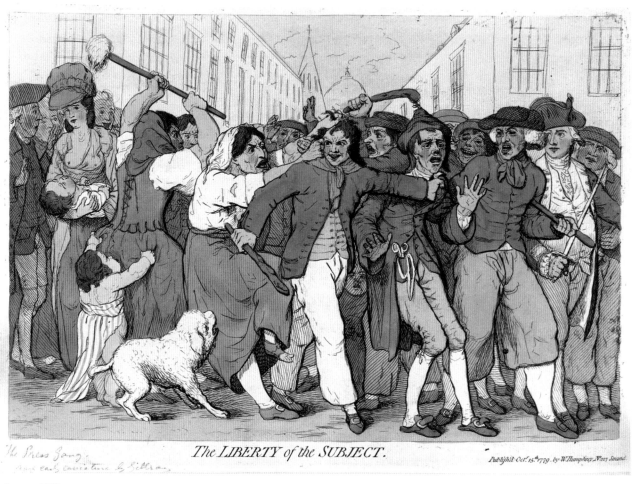

The LIBERTY of the SUBJECT.

James Gillray
The Liberty of the Subject
William Humphrey, 15 October 1779.
Hand-coloured etching.
NMM PAG8527

The press gang was a common subject for caricaturists wishing to highlight the state's coercive authority, and one that chimed particularly well with Whiggish ideas about the rights of the individual. In *The Liberty of the Subject*, a rare early work by Gillray, a London tailor is seized by a press gang on a City street, despite the furious resistance of an angry crowd. The print was produced during the War of American Independence, in which Britain fought against a new and potentially subversive nation inspired by novel democratic ideals. Debates over the meaning of 'liberty'

became widespread as the American colonists drew considerable sympathy from radicals across the country. In this image, the notion of 'British liberty' is undermined as a tailor, wearing a green coat and with a pair of scissors in his pocket, is forcibly carried off to sea to fight the American rebels. However, whether deliberately or not, the image misrepresents how the press gang operated. The tailor's trade was considered at the time to be a passive, even effeminate occupation, making the gang's terrified victim an unlikely recruit for the rigours of naval service.

Politics and the Navy

John Barlow after Samuel Collings

Manning the Navy

Bentley & Co,
1 June 1790.
Hand-coloured etching with engraving.
NMM PAD4732

The press gang became the most infamous of all the navy's activities. *Manning the Navy*, has sailors using cudgels to coerce individuals into the navy. A distressed wife looks on in horror, while a man is forced to his knees by the party of sailors. The print also suggests the gang's indiscriminate tendencies; on the right, a portly individual is gripped by a sailor, while a grinning officer inspects him.

MANNING THE NAVY.

THE USE of a GENTLEMAN, or Patronage for the Admiralty.

Charles Williams

The Use of a Gentleman – or Patronage for the Admiralty

Thomas Tegg,
c1810.
Hand-coloured etching.
NMM PAG8600

The press gang's undiscerning nature was a frequent presence in eighteenth-century caricature, though it served comic as well as political roles. In this humorous take on impressment, the artist turns the tables and imagines a press gang bringing in members of the social elite. Two men attired in expensive and dandified clothing are marshalled to the Admiralty. One pleads that he is indeed a gentleman, and well known in Bond Street. The officer is unbowed, and hopes they will be able to teach his crew some manners: 'Yes, yes! I see you are every inch a Gentlemen, and just the person we want, my men have pressed a d—'d numer [*sic*] of Blackguards, and we want a Gentlemen on board to teach them good manners!' The pointless and shallow occupations of the upper classes are painted in colourful contrast to the straight-talking, jolly seamen fighting for Britain. On the far left, a sailor looks forward in mock delight to his future lessons in etiquette.

When, in April 1797, the sailors of the Channel Fleet anchored at Spithead mutinied, the nation at large was shocked by the news. Having come to rely on and trust the navy as its main protector, Britain now found its principal arm of defence against revolutionary France rendered impotent. In reality, the Spithead mutiny was little more than a workplace dispute. The paucity of naval wages, unchanged for over a century and made worse by inflation, was acknowledged as a legitimate issue, and for the most part, the mutineers maintained a respect for onboard discipline, conducting themselves with dignity. The Admiralty conceded to the mutineers' demands on 15 May 1797, and two days later the fleet resumed its station off the coast of France. However, the following weeks gave rise to two more mutinies, at the Nore at the mouth of the Thames, and at Yarmouth. Voicing more

Isaac Cruikshank

The Delegates in Council or Beggars on Horseback

Samuel Fores, 9 June 1797.

Hand-coloured etching.

NMM PAG8535

The radical undertones of the mutinies, though prone to exaggeration, proved irresistible for caricaturists. *The Delegates in Council* presents a menacing view of the ringleaders sitting at a cabin table while Admiral Howe, who had negotiated with the mutineers at Spithead, stands in apparent deference. The delegates hold guns and swords, while in the background a portrait of Britannia hangs upside down as a graphic symbol of the social order being turned on its head. Underneath the table, Fox and his fellow Whigs surreptitiously encourage the mutineers, indicating their complicity, and suggesting that the

mutineers were pawns in a more sinister political game.

The Delegates in Council appeared two weeks after the resolution of the Spithead mutiny, but while events at the Nore were approaching a bitter, self-destructive climax as the mutinous crews attempted to blockade the Thames. Cruikshank's representation of the British sailor contrasts sharply with the bawdy, comical figure of fun more often promoted by caricaturists. The assembled seamen take on a dangerous appearance, threatening to subvert the social and political make-up of the navy.

extreme demands, these latter incidents proved to be more divisive among the sailors involved, and reached a bloodier conclusion. Starved of supplies and well-founded support, the mutinies collapsed and twenty-nine ringleaders, nominally led by Richard Parker, were executed on board their own ships.

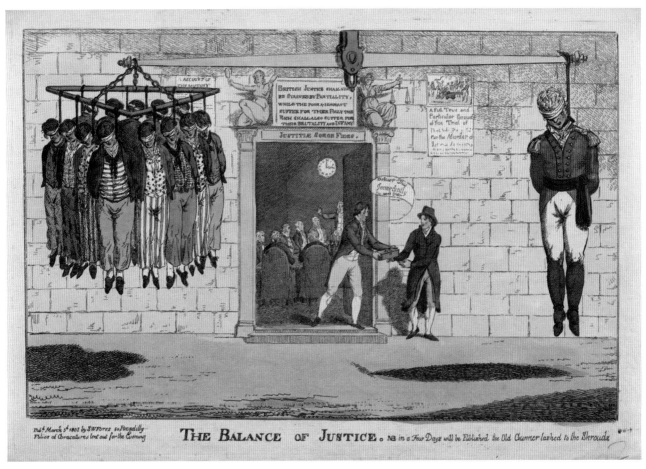

The Balance of Justice
Samuel Fores,
3 March 1802.
Hand-coloured etching.
NMM PAF3882

When a Royal Navy squadron stationed at Bantry Bay off the southwest coast of Ireland refused to sail to the West Indies in December 1800, punishment was swift and forceful. Fourteen men identified as ringleaders were tried for mutiny, and six were hanged on board their ships at Spithead as an example to the rest of the fleet. By coincidence, Joseph Wall, the former governor of the slave-trading centre at Goree, West Africa, was at the same time on trial for the murder of a sergeant who had died after a particularly vicious flogging ordered by Wall. After repeated attempts to obtain a pardon, he was executed on 28 January, two weeks after the Bantry Bay mutineers.

The unnamed artist behind the print seems to suggest that no one, regardless of rank, could escape justice. However, by implying that the life of a single officer was worth a dozen or more sailors, the print also points to a growing sense of *in*justice at the disproportionate punishments levelled against those serving at sea. Both incidents were reported in detail, and while the repentant Bantry Bay mutineers (many of whom had served throughout the war) elicited some sympathy after their

trial, the large crowd who had gathered outside Newgate prison to see Wall die showed little compassion for the disgraced army officer who had come to embody the brutality of the British military to its own men. A contemporary account of Wall's much publicised crimes is posted on the wall behind, next to the figure of Justice whose emblematic scales echo the over-sized balance from which both condemned parties hang. However, *The Balance of Justice* observes but does not judge the proceedings, focusing instead on the delicate balancing act performed by the state when dealing with crime in the armed forces.

Such brutal imagery, though entirely in keeping with the extreme violence of its subject, is unusual in the realm of caricature. The image of fourteen men hanging by their necks is starkly devoid of humour, and reminiscent of Jacques Callot's earlier and dark reflection on war crimes committed against civilians during the Thirty Years War, but here the printmaker's characteristic style and the introduction of colour gives *The Balance of Justice* a disconcertingly light-hearted appearance.

*J*ohn Bull embracing the Pie-Man and The Yankey Torpedo show how caricaturists continued to engage with naval politics in complex and contrasting ways in the years after the Battle of Trafalgar. However, the end of hostilities against France in 1815 fundamentally altered the military preoccupations of Britain's printmakers. Gillray, perhaps the artist most capable of exploiting the satirical potential of the navy, died in June 1815, two weeks before Napoleon's final defeat at Waterloo. These two events marked the end of an era for British graphic satire – an end to the profusion of comic

John Bull embracing the Pie-Man, or a Friendly Visit to Zeland
Thomas Tegg, 15 October 1807.
Hand-coloured etching.
NMM PAF4002

In 1807 Denmark remained neutral in the war between Britain and France, but the Danish fleet of twenty-two ships of the line loomed large in the power calculations of both protagonists: Napoleon wanted to secure it for France, and Britain needed to stop the enemy having it. The British Foreign Secretary, George Canning, attempted a diplomatic solution, but the Danes were not persuaded by Canning's argument that their national interests were more closely aligned to those of Britain. Denmark refused to merely hand over its fleet. Under the pretext of secret intelligence that suggested a Danish-French alliance, Britain launched a pre-emptive attack against the city of Copenhagen. The bombardment of the city succeeded in securing the Danish fleet, but at the cost of many civilian lives. Most of Europe was outraged by this aggressive act, and Britain's political elite were divided. While some, including William Cobbett and William Wilberforce, defended the attack as a necessary act of self-defence, Lord

Erskine, a leading Whig, was adamant in his opposition, famously declaring that 'if hell did not exist before, Providence would create it now to punish ministers for that damnable measure'.[14]

John Bull embracing the Pie-Man is unusually vitriolic in its critique of the operation, drawing an analogy with the popular eighteenth-century nursery rhyme *Simple Simon*. John Bull, the embodiment of the British character, assumes the role of the hapless Simon with an unexpected, aggressive turn, as he insists that his actions are born 'out of pure Friendship' – to which the Dane replies, 'Master Bull you will throttle me with kindness, you have upset all my pies.' John Bull's behaviour underscores the violence of the British ships that bombard the burning city in the background. The result is a dark satire and an excellent example, not only of how caricature responded to contemporary events, but also how it could engage with public opinion even when directed against the policy and actions of the navy.

The YANKEY TORPEDO. *Pub. Novr 1st 1813 by Thos Tegg in Cheapside*

William Elmes

The Yankey Torpedo

Thomas Tegg

1 November 1813.

Hand-coloured etching.

NMM PAF4164

Thomas Tegg's cruder tastes combine with transatlantic events and an eye for the macabre in the *The Yankey Torpedo*, in which Jack Tar presents his posterior to an American cannon-spewing sea monster and invites his enemy to 'kiss my—tafferal'. William Elmes's imaginative and unusually violent caricature was inspired by a new invention – the torpedo, devised by the American engineer Robert Fulton and used to powerful effect in 1813 during the defence of New York against the British blockading fleet. An American sloop was filled with gunpowder and used as a primitive torpedo, ignited by clockwork. Elmes represents the new threat of Fulton's invention, which caused great indignation in Britain by undermining the accepted rules of naval engagement, merging the ordnance of conventional warfare with the unquantifiable terror of the mythological sea monster, the Devil, and Death.

and satirical prints that had championed, lampooned and critiqued the Royal Navy during the previous half century of war. The British love of graphic satire remained, but while the likes of Ackermann, Fores and Tegg enjoyed continued success well into the next decade, the twin naval icons of Jack Tar and Nelson became less prominent in their shop windows as the navy's role in Britain's domestic and international affairs became less visible. For the next generation of artists, the possibilities of caricature were transformed by new and industrial methods of printmaking and distribution and, as far as the navy was concerned, by the changing character and remit of Britain's armed forces. Thus, the topical satires that accompanied the navy's next great trial, the Crimean War in the 1850s, took on a very different appearance – defined, above all, by the phenomenally successful *Punch* magazine, and shaped by a navy that was itself much altered in composition and capability.

Further Reading

Few books have addressed specifically the representation of the navy in caricature. Notable exceptions include Alexandra Franklin and Mark Philp, *Napoleon and the Invasion of Britain*, Oxford, 2003; and the related volume of essays, edited by Philp, *Resisting Napoleon: The British Response to the Threat of Invasion, 1797–1815*, Farnham, 2006. There have been a number of books that consider the broader political and cultural resonance of the navy in eighteenth-century Britain: see in particular Margarette Lincoln, *Representing the Royal Navy: British Sea Power, 1750–1815*, Aldershot, 2002; and Timothy Jenks, *Naval Engagements: Patriotism, Cultural Politics, and the Royal Navy 1793–1815*, Oxford, 2006. There is a vast amount of scholarship that covers the history of the eighteenth-century Royal Navy: the best overview is N A M Rodger, *The Command of the Ocean: A Naval History of Britain 1649–1815*, London, 2004. The character and significance of Horatio Nelson is adeptly treated in Roger Knight's *The Pursuit of Victory: The Life and Achievement of Horatio Nelson*, London, 2005; and Andrew Lambert's *Nelson: Britannia's God of War*, London, 2004. For work on 'Jack Tar' as a British icon, see Isaac Land, *War, Nationalism and the British Sailor, 1750–1850*, London, 2009; and Geoff Quilley, *Empire to Nation: Art, History and the Visualization of Maritime Britain, 1668–1829*, London, 2011.

There is a burgeoning literature that considers British caricature, though much of the recent work has focused on the work of James Gillray, the only satirist of the period also to have been the subject of a major retrospective exhibition; see Richard Godfrey, ed, *Gillray and the Art of Caricature*, Tate exhibition catalogue, London, 2001. Two publications provide a fascinating view of some of the earliest responses to Gillray's work: *The Caricatures of Gillray; with Historical and Political Illustrations, and Compendious Biographical Anecdotes and Notices*, London and Edinburgh, 1818; and Christiane Banerji and Diana Donald, eds, *Gillray Observed: The Earliest Account of his Caricatures in London und Paris*, Cambridge, 2009. The best work on the subject more generally includes: Diana Donald, *The Age of Caricature: Satirical Prints in the Reign of George III*, New Haven and London, 1997; Tamara L Hunt, *Defining John Bull: Political Caricature and National Identity in Late Georgian England*, Aldershot, 2003; Vic Gattrell, *City of Laughter: Sex and Satire in Eighteenth-century London*, London, 2006; and Todd Porterfield, ed, *The Efflorescence of Caricature, 1759–1838*, London, 2010. For a longer view of British caricature and graphic satire, see M Dorothy George, *Hogarth to Cruikshank: Social Change in Graphic Satire*, New York, 1967; and Martin Myrone, *Rude Britannia: British Comic Art,* Tate exhibition catalogue, London, 2010.

NOTES

1   M T S Raimbach, *Memoirs and Recollections of the Late Abraham Raimback, Esq, Engraver* (London, 1843), p105.

2   Johann Christian Hüttner, writing for the German journal *London und Paris* in 1806, cited in Christiane Banerji and Diana Donald (trans and eds), *Gillray Observed: The Earliest Account of his Caricatures in London und Paris* (Cambridge: Cambridge University Press, 1999), p247.

3   *The Caricatures of Gillray; with Historical and Political Illustrations* (London: John Miller, 1818), pp26–7.

4   *A Tragic-Comic Dialogue, Between the Ghost of an A——l, and the Substance of a G——l* (London, 1759).

5   Emma Hamilton to Horatio Nelson, 8 September 1798, British Library, London, Add MSS 34989, ff4–7.

6   A B Paget, (ed), *The Paget Papers* (London, 1896), vol 2, p162.

7   Suffolk Record Office, Ipswich, SA 3/1/2/1, 2 February 1798.

8   Napoleon I, *Correspondence de Napoleon Premier* (Paris, 1858–69; 32 vols) vol 11, p514.

9   Alan and Veronica Palmer, (eds), *A Dictionary of Historical Quotations* (London, 1991), p232.

10 Leslie Mitchell, *Charles James Fox* (Oxford: Oxford University Press, 1992), p205.

11 Richard Brinsley Sheridan, *Sheridan's Address to the People: Our King! Our Country! And our God!* (London, 1803).

12 William Shakespeare, *Macbeth*, Act 3, Scene 4.

13 *Naval Chronicle*, 22 (1809), p19.

14 Henrietta Ponsonby, Lady Bessborough to Granville Leveson-Gower, 18 December 1807. Castalia Countess Granville, (ed), *Lord Granville Leveson Gower, Private Correspondence, 1781 to 1821* (London, 1916), p315.

ACKNOWLEDGEMENTS

In writing this book the authors have accumulated many debts of gratitude. The majority of the caricatures illustrated come from the NMM's collection, and several are published here for the first time. To make this possible, the prints were expertly conserved by Clara de la Pena and beautifully photographed by Tina Warner and David Westwood. We also owe considerable thanks to Emma Lefley and Doug McCarthy in the Picture Library, and also to Josh Akin, Lucinda Blaser and Will Punter. We would also like to thank our curatorial colleagues for their support in bringing this project to fruition, and a number of other individuals who have offered advice or assistance during the project: Tarah Butler, Roger Knight, Emily Mann and Lauren Wolper. We are particularly indebted to Pieter van der Merwe and Catherine Watson who read and commented on the entire manuscript. This book could not have been written without the encouragement and publishing nous of Rebecca Nuotio, and the organisational talents of Kara Green. Finally, we would like to thank Julian Mannering, Stephanie Rudgard-Redsell and Steve Dent at Seaforth publishing, who have guided the book through to publication with impressive efficiency.

This publication accompanies an exhibition at the National Maritime Museum, open between October 2012 and February 2013.

Dr James Davey is Curator of Naval History at the National Maritime Museum.

Dr Richard Johns is Curator of Prints and Drawings at the National Maritime Museum.

First published in Great Britain in 2012 by
Seaforth Publishing,
Pen & Sword Books Ltd,
47 Church Street,
Barnsley S70 2AS

www.seaforthpublishing.com

British Library Cataloguing in Publication Data
A catalogue record for this book is available from the British Library

ISBN 978 1 84832 146 5

Typeset and designed by Stephen Dent
Printed and bound in Great Britain by Henry Ling Ltd.

Front cover: **Detail of** *John Bull taking a Luncheon*, 1798, James Gillray, NMM PAF3941 (see page 8).

Back cover: **Detail of** *A Land Storm*, c1815, Thomas Rowlandson. NMM PAF3829

Inside covers: *Portsmouth Point*, 1811, Thomas Rowlandson, NMM PAF3841 (see page 39).

Broadsides explores the history of the Royal Navy during the second half of the eighteenth century through the lens of contemporary caricature. This book presents a unique perspective on the navy's place in British society and culture through the combined historical and art-historical expertise of the authors. Beautifully illustrated throughout, *Broadsides* includes well-known images and a selection of previously unseen prints from the National Maritime Museum's collection, ranging from politically calculating satire to bawdy representations of the drunken sailor, from artists including Gillray, Rowlandson and Cruikshank. With contextual information alongside detailed analysis of individual prints, this book sheds new light on some of the most celebrated images of the age.

Dr James Davey is Curator of Naval History
at the National Maritime Museum.

Dr Richard Johns is Curator of Prints and Drawings
at the National Maritime Museum.

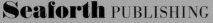

Seaforth PUBLISHING
in association with

NATIONAL
MARITIME
MUSEUM

NAPOLEONIC WARS
ISBN 978-1-84832-146-

£16.99

9 781848 321465

www.seaforthpublishing.com